An Introduction to

Animal Behaviour

Second Edition

Aubrey Manning
B.Sc., D.Phil.

Reader in Zoology, University of Edinburgh

Edward Arnold

First published 1967
by Edward Arnold (Publishers) Ltd.,
25 Hill Street,
London, W1X 8LL

Reprinted 1968
Reprinted 1969
Reprinted 1970
Reprinted 1971
Second edition 1972
Reprinted 1973

Boards edition ISBN: 0 7131 2360 5
Paper edition ISBN: 0 7131 2361 3

Printed in Great Britain by
William Clowes & Sons, Limited, London, Beccles and Colchester

A series of student texts in

CONTEMPORARY BIOLOGY

General Editors:
Professor E. J. W. Barrington, F.R.S.
Professor Arthur J. Willis

Preface to the Second Edition

The basic plan of the book remains unchanged but in preparing this second edition I have taken the opportunity to update and extend the discussion at a number of points. Inevitably the text is longer, but I hope it is still compact enough for introductory courses in animal behaviour.

I had the good fortune to receive constructive comments on the first edition from a number of people and I have tried to meet their criticisms. Every chapter, except the first, contains some parts of extensive re-writing where I felt new information or a lack of clarity in the original demanded it. Perhaps the most extensive changes are in Chapter 4, which now contains an extended discussion on the nature of aggression. Dispute on this topic has become widespread, and it offers a good opportunity to illustrate some of the problems inherent in the study of motivation.

The study of animal behaviour continues to expand rapidly within diverse disciplines. It seems more than ever worth while to introduce the subject through an approach based upon general biological principles. In particular, work on animal societies is increasingly coming to involve sociologists and anthropologists as well as zoologists. The new Chapter 10 attempts to fill an obvious gap in the first edition and to put some of the recent work on social organizations, which has yielded such fascinating results, into this biological perspective. In preparing this chapter, I have been greatly assisted by the comments and criticisms of my colleague John Deag.

My book was originally prepared without the benefit of Hinde's masterly survey of the field. This revision has certainly benefited greatly from access to his second edition. It will have served its purpose if it leads readers on to explore his advanced text for themselves.

Edinburgh
1972

A.M.

Preface to the First Edition

The study of animal behaviour is now becoming recognized as one of the major branches of biology. Not only does it serve to draw together disciplines as diverse as neurophysiology and ecology, which are otherwise studied in isolation, but it also offers problems of its own as intellectually challenging as any found in science.

There are, at present, few textbooks which approach behaviour from a biological point of view and link it to physiology on the one hand and psychology on the other. This book is an attempt in this direction. It is a personal selection of topics which I consider to be important, and makes no claim to cover the whole field—this is scarcely possible in a book of this length. I have tried to cut across any classification of behaviour by function—feeding, fighting, courtship, etc.—and have chosen topics which help in analysing the organization of behaviour within the individual animal. I have not attempted to make a thorough comparative survey through the animal kingdom. This is quite unprofitable at present, because our knowledge is very patchy and most topics are best exemplified from particular groups.

The book assumes some basic knowledge of physiology and genetics, but no more than would be provided in a school course. I hope the book will be useful for undergraduate courses and for school work. I hope, too, that it will live up to its title and serve as an *introduction* to the whole study of behaviour; there are fairly extensive references which should enable anyone to delve deeper where their interest is caught. I have tried to strike a balance between references to detailed, experimental papers and those to more general reviews of the literature.

It is a pleasure to acknowledge my indebtedness to Dr. Margaret Bastock, Dr. Philip Guiton, Mr. Peter Slater and Dr. David Wood-Gush, all of whom read parts of the typescript and helped greatly by their comments. I am also grateful to the various authors and publishers who gave permission to use their material for illustrations; they are acknowledged individually in the captions. Finally I must thank Dr. Peter Brunet, who was General Editor of the series during the preparation of this book, for all his help over the past 18 months.

Edinburgh   A.M.
1967

Contents

I

Reflexes and Complex Behaviour

There are two main approaches to the study of behaviour, the physiological and the psychological. Physiologists are primarily interested in mechanisms, and aim at giving an explanation of behaviour in terms of the functioning of the nervous system. Psychologists are more concerned with the behaviour itself, studying those factors in an animal's environment and history which affect the development and performance of overt behaviour.

These two approaches are both essential and complementary. Physiologists sometimes like to emphasize that their methods are the more fundamental, and it is true that ultimately we shall hope to explain behaviour in terms of the functioning of the basic units of the nervous system—the neurons. However, the main function of the nervous system is to produce behaviour and we must investigate the end product in its own right. Many of the most important aspects of neural organization can at present be expressed only in behavioural terms.

This book does not itself contain much formal physiology, but any study of behaviour which is not mindful of physiology is very unrealistic. Indeed one of the most flourishing areas of psychological research at present is that called 'physiological psychology', in which attention is focused simultaneously on an animal's behaviour and its associated physiology. One of the obstacles to closer links between physiologists and other behaviour workers is the small overlap between the types of behaviour each group studies. Complete physiological analysis is not yet possible save in the simplest of responses involving only a handful of neurons. This opening chapter will examine some possible ways of bridging the gap between such simple responses and the more complex ones which will be our main concern.

'Behaviour' includes all those processes by which an animal senses the external world and the internal state of its body, and responds to changes which it perceives. Many of such processes will take place 'inside' the nervous system and may not be directly observable. What the animal does may involve violent activity or complete inactivity, but all are equally behaviour. Even if amount of movement is not important, we might tend to exclude some movements from a behavioural study because they are too simple. Most of this book will be concerned with animals performing rather complex patterns of movement; building a nest, displaying in front of a mate, running through a maze to get at food and so on. But how simple can a movement be and still count as behaviour? Few would count the jerk of our leg when the tendon below our knee-cap is tapped. What about the movement which closes our eyes when something flashes towards them, or our reaction to stepping barefoot on a sharp object? Reflex actions of this type are usually considered to be rather distinct from complex behaviour and in many universities they are discussed by a physiologist at the tail end of a course on the nervous system. 'Behaviour' is taken as a separate topic, often taught by a different person, and the foundations of a false dichotomy are laid down.

It is not possible to draw a firm line between reflexes and complex behaviour. Neither term can even be defined satisfactorily, but common sense will determine their use in what follows. Clearly complex behaviour can incorporate many reflexes; the swallowing reflex is the culmination of complex food-seeking behaviour, and the reflexes controlling balance and walking are involved in almost all complex behaviour. The two represent the ends of a continuous scale and somewhere intermediate along this scale are the 'tropisms' by which animals orientate themselves with respect to light, gravity and other environmental factors (see Fraenkel and Gunn[74]). Tropisms have some characteristics of reflexes, but they involve movements of the whole body and not just a single group of muscles. Whilst we can often describe the neural pathways and properties of a reflex in a rather exact manner, we can scarcely ever do this for complex behaviour. In the latter, the neurons involved must be numbered at least in hundreds and the impinging variables become legion. Nevertheless we should not be put off by quantitative differences no matter how great. There is much to be learnt from a comparison between the properties of reflexes and more complex behaviour, and it is a useful way to give some physiological foundations to a study of behaviour.

In 1906 C. S. Sherrington's book *The Integrative Action of the Nervous System*[241] was published. Sherrington, more than any other single person, can be regarded as the founder of modern neurophysiology. In his book he considered the way in which reflexes operate and how the central nervous system integrates them into adaptive behaviour, combining information gathered from different sources, arranging sequences of action and allocating priorities. *Integrative Action* is a scientific classic which can still be read for pleasure and profit. Sherrington had to work with apparatus we now regard as crude. There were no electronic stimulators or oscilloscopes, simply induction coils and levers, attached to the limb of the animal and writing on the smoked drum of a kymograph.

In the first few chapters of his book Sherrington discusses some of the properties of reflexes and contrasts them with those of the same movements when elicited by direct stimulation of nerves to the muscles concerned. We may, in turn, use part of his classification to compare the properties of reflexes and more complex behaviour.

Latency

Reflexes and complex behaviour both show latency in response—there is a delay between giving a stimulus and seeing its effect. Sherrington calculates that, allowing time for conduction along axons, the dog flexion reflex in which it withdraws its leg in response to painful stimuli on the skin, should have a latency of about 27 msec. In fact the latency usually lies between 60 and 200 msec.

Latencies are harder to measure for complex behaviour because there are often difficulties in fixing precisely the time of onset of the stimulus, but they are none the less vivid. Wells[279] describes how, when a tiny shrimp (*Mysis*) is presented to a newly-hatched cuttle-fish (*Sepia*), there is no detectable response for perhaps as long as 2 minutes. Then the nearest eye of the cuttle-fish turns to fixate on the shrimp. There is a further delay, but usually only a few seconds, before the cuttle-fish turns towards the shrimp so that both of its eyes are brought to bear. Another brief delay follows, and then it launches an attack and seizes the shrimp with its tentacles. A male fruit-fly (*Drosophila*) is often not aroused by its first encounter with a female. He may touch her with his fore-legs, but stands still if she moves away. Only at a subsequent encounter some seconds later does he begin his courtship display.

The latency of reflexes is known to be due to delay in the transmission of impulses across the synapses (a term we owe to Sherrington) between one neuron and the next. It is hardly surprising to find delays between stimulus and response in complex behaviour for in the chain between receptors and effectors there must often be dozens of synapses to cross.

With reflexes it is found that the stronger the stimulus, the shorter the latency (see Fig. 1.1), but measurements of the latency of complex behaviour are inextricably intermingled with factors affecting thresholds. Reflexes have a relatively stable and rather low threshold. They are one of the body's protective mechanisms and need to be constantly on tap if required. With complex behaviour, in contrast, thresholds vary enormously. A food stimulus which produces an immediate and intense response from a hungry dog may be ignored an hour later after it has fed. Consequently any attempt to investigate how latencies change with stimulus strength is hedged about with difficulties. There are two ways to get an estimate of latency in such cases. One can try to get all one's subjects equally motivated, i.e. equally hungry if investigating latency to a food stimulus. Secondly, one can use a large number of animals with varying thresholds and then compare statistically the range of latencies given to one stimulus with that given to another. This method was used by Hinde[111] who measured the latency between presenting various frightening stimuli to chaffinches and their first alarm call. Just as with reflexes, the stimulus known on other grounds to be strongest, produced the shortest latency.

After-discharge

When a motor nerve is stimulated directly the muscle it supplies contracts and then relaxes within a few seconds of the stimulus ceasing. The

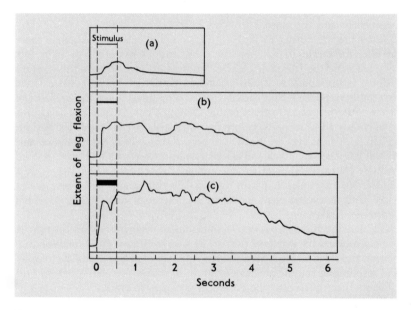

Fig. 1.1   The extent and persistence of the dog's flexion reflex with three strengths of stimulus, each of 0·5 sec duration. The area enclosed between the line and the X axis gives a measure of the 'amount' of the response. Even with a weak stimulus (a) the after-discharge represents 75% of the total 'amount'; with the strongest stimulus (c) it represents over 90%. Note that the latency of the response decreases as the stimulus increases in strength. (Modified from Sherrington,[241] 1906, *The Integrative Action of the Nervous System.* Charles Scribner's Sons, New York.)

same is not true for muscle contractions elicited through a reflex arc; they often persist at full intensity for many seconds after the stimulus ends and then relax gradually. Sherrington found that the length and strength of this after-discharge in the dog's flexion reflex was related to the strength of the stimulus (see Fig. 1.1).

After-discharge is a familiar phenomenon in complex behaviour. Hinde[107] studied the 'mobbing' response which chaffinches make towards owls and other predators. The bird comes quite close to the owl and appears restless, making frequent short flights sometimes towards, sometimes away from it. Repeatedly the chaffinch gives the 'chink' alarm call. In the wild this behaviour attracts other birds who also begin mobbing until the owl is surrounded by a flock of harrying small birds and it often beats a retreat. Hinde found that the rate of calling 'chink' was a good measure of the strength of this mobbing response in an individual chaffinch alone in a large cage and found that calling persisted long after the owl was removed. The stronger the chaffinch was calling at the time, the longer was this after-discharge.

A male *Drosophila* who is courting an unreceptive female sometimes loses contact with her if she jumps or flies away. In such instances he often continues courtship movements for several seconds, orientating himself to the spot where the female was last perceived. With complex behaviour such as this it is difficult to state categorically that all stimuli are removed. When Sherrington switched off his induction coil he could then be certain that no stimulus relevant to the reflex was reaching the dog. But perhaps after she has gone, some of the female *Drosophila*'s smell remains behind and the male's continued courtship is not strictly an after-discharge. Again, Hinde could show that whilst the owl was in the cage the chaffinch associated it with other objects close by. Some of the apparent after-discharge may have been a response to these objects which remained after the owl was gone. In spite of these qualifications there can be little doubt that complex behaviour does show what Lorenz[157] aptly called 'reaction momentum'; it takes time to slow down once aroused. Prolonged excitability at synapses must be involved just as it is in reflex after-discharge. In Chapter 3 we shall discuss how a stimulus can 'arouse' an animal and thus, not only do its effects persist, but the animal may become more responsive to subsequent stimuli.

Summation

One of the most obvious integrating properties of the central nervous system is its ability to summate stimuli coming at different times (temporal summation) and from different places (spatial summation). Sherrington gives several beautifully clear examples. The scratch reflex of the dog is elicited by irritating stimuli anywhere on a saddle-shaped area of its back. The hind leg on the same side is brought forward and rhythmically scratches at the spot. With weak stimuli, a series of 5 or 10 given in rapid succession may not evoke any response, but after 20 or 30 scratching appears—the stimuli have been summed in time. Fig. 1.2 shows the spatial summation of stimuli from two areas of skin 8 cm apart; neither is strong enough alone to provoke scratching, but they are effective when given together. Sherrington found that points up to 20 cm apart can summate in this way, but the effect diminishes with distance up to this maximum.

Dethier[62] studied the stimuli which cause blow-flies to extend their proboscis preparatory to drinking. The flies can detect sugars and other food substances with sensory hairs on their fore-tarsi. They search for food by running over a surface and extending their proboscis when the front legs encounter anything suitable. As measured by this proboscis extension, the flies can detect sugars at very low concentrations. Dethier found that when only one leg is dipped into a solution the lowest concentration of sucrose to

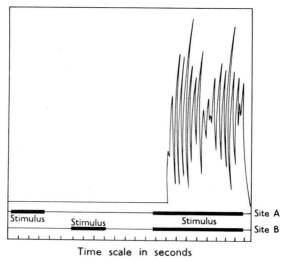

Fig. 1.2 Spatial summation leading to the appearance of the scratch reflex in the dog. The tracing represents the movements of the dog's leg when scratching. A and B are two points on the shoulder skin and weak stimuli given singly, first at B, then at A, do not evoke the reflex. When both points are stimulated simultaneously the reflex appears with a latency of about 1 sec. (Modified from Sherrington,[241] 1906, *The Integrative Action of the Nervous System.* Charles Scribner's Sons, New York.)

which 50% of flies respond is 0·0037M. However, if both legs are stimulated together there is summation and now 50% of flies respond to 0·0018M. He further found that if one leg was stimulated by unpleasant substances such as salt or acid which tend to make a fly retract its proboscis, whilst simultaneously the other leg is dipped into sugar, a kind of algebraic summation occurs. The stronger the acid on one side, the stronger the sugar has to be on the other to overcome the negative effects and cause a fly to extend its proboscis.

With more complex behaviour summation frequently occurs between stimuli of quite different types perceived by different sense organs. We all know how the sight and smell of food summate when we are hungry. Beach[22] showed that male rats respond sexually to a combination of olfactory, visual and tactile stimuli from a receptive female. Young males do not respond unless two such sources are available—it does not matter which two. Mature males, with previous sexual experience, will respond to one type of stimulus alone.

'Warm-up'

Sherrington found that some reflexes do not first appear at full strength but, with no change to the stimulus, their intensity increases over a few seconds. Hinde's chaffinches show a similar type of 'warm-up' effect. Fig. 1.3 is a record of the rate of calling by a chaffinch in successive 10-second periods after presenting an owl and the maximum rate is not reached for about $2\frac{1}{2}$ minutes. Sevenster-Bol[238] counted the courtship movements made by male sticklebacks to a female confined in a glass tube (see p. 82 for more details of this experiment). She found that over a 5-minute period the number of movements rose steadily in each successive minute of the period. Sherrington was able to show that 'warm-up' in some reflexes is due to the summation of stimuli which come to evoke a response from more and more motor nerve fibres, producing a stronger contraction. He called this phenomenon 'motor recruitment'.

Some analogous process probably occurs with complex behaviour, but we commonly see that it is not only the intensity of a response which changes with persistent stimulation; its nature changes as well. Again, in some later work, Sherrington[242] provides an excellent example from what he calls the cat's 'pinna reflex'. Repeated tactile stimuli to a cat's ear first cause it to be laid back. If stimulation persists the ear is fluttered, thirdly the cat shakes its head and when all else fails to remove the irritation it brings up its hind leg and scratches. Clearly there is more involved here than the recruitment of extra motor nerve fibres. Mechanisms which control patterns of movement such as ear-fluttering and head-shaking must be recruited. Perhaps all these mechanisms are activated in some way by the stimuli to the ear but their thresholds are different. That for laying back the ear will have the lowest threshold, with successively higher ones for the other three patterns. As the persistent stimuli summate, and the threshold for activating each pattern is passed, it replaces the pattern which preceded it. Workers studying complex behaviour frequently rank the various patterns they observe on a similar 'intensity scale' of increasing thresholds. A system similar to that suggested for the cat has been used by Bastock and Manning[19] to explain the fact that a male *Drosophila* switches from one courtship pattern to another when courting a female whose behaviour remains constant.

Fatigue

When a muscle is made to contract by regularly repeated stimulation of its motor nerve it continues to respond for a very long time—several hours or more. This is not the case when the same muscle is stimulated through a reflex arc. Sherrington found that the dog scratch reflex begins to wane after about 20 seconds of continuous mechanical or electrical stimulation at a single point on the skin. The movements of the leg become weaker and lose their rhythm. This fatigue is certainly not due to the muscles because if the flexion reflex which uses the same muscles is now evoked, it appears

Fig. 1.3 The 'warm-up' and subsequent 'fatigue' of alarm calling when a chaffinch is presented with a stuffed owl in its cage. The maximum rate of calling occurs after about $2\frac{1}{2}$ min. (From Hinde,[107] 1954, *Proc. R. Soc., B*, **142**, 306.)

at full strength. Further, if the site of stimulation on the skin is switched a few centimetres the scratch reflex recovers and, after a short rest, the first site once more becomes effective when stimulated (see Fig. 1.4).

This means that the origin of the fatigue must lie somewhere between the sense organs in the skin and the origin of the motor nerve. Sherrington believed that it was caused by increased resistance to transmission across the synapses between the internuncial neurons, which convey impulses down the spinal cord, and the motor neurons.

Franzisket[76] has confirmed that the internuncials are the site of fatigue in the leg-wiping reflex of the frog. When its back skin is touched lightly the frog brings up its hind leg and makes a single wiping movement. This fatigues after some dozens of stimuli but, as with the dog, shifting the site of stimulation will keep the leg working. Franzisket observed that tiny twitches of the frog's skin around the point where it was touched continue after the leg has ceased to respond. This shows that the sensory nerves from the skin are still working and impulses are being transmitted to the motor nerves which supply the local muscles of the skin. The leg-wipe has disappeared because the internuncial neurons linking the sensory nerves to those of the leg muscles are not transmitting.

Fatigue is always a feature of complex behaviour. Compare Fig. 1.3, which shows how the calling rate of the chaffinch declines in the mobbing situation, with the left-hand portion of Fig. 1.4, showing the decline of the

Fig. 1.4   The 'fatigue' of the dog's scratch reflex. Prolonged stimulation at site A eventually ceases to be effective, but stimulation at site B immediately after brings the response back and after only 9 sec further stimulation at A evokes the response again. (Modified from Sherrington,[241] 1906, *The Integrative Action of the Nervous System*. Charles Scribner's Sons, New York.)

dog's scratch reflex. In the chaffinch example, however, 'fatigue' is merely a name for what we observe—that the response declines even though the stimulus is maintained. Taken alone, such observations tell us nothing about the mechanisms involved and certainly these are not the same in every case.

Sometimes, as with reflexes, a change in stimulus brings about a re-awakening of a response which has fatigued. Young passerine birds beg for food when the parent bird arrives at the nest by stretching up their necks and gaping with wide-open beaks. Prechtl[207] found that any dark object appearing over the rim of the nest evokes begging and so does a jolt to the nest, which normally signals the parent alighting. If begging is repeatedly evoked by visual stimuli it becomes weaker and eventually there is no further response. If the nest is now jolted begging appears at full strength again. In such a case we may call the original fatigue 'stimulus-specific', i.e. other types of stimulus will overcome it.

Hinde's[108] work with the chaffinch mobbing response shows that there are other types of fatigue which we can describe in behavioural terms. Fig. 1.5 shows the results of one of his experiments. An owl is presented to a bird for 30 minutes; by the end of this time it has usually stopped mobbing and as a measure of the strength of its response Hinde counts the number of 'chink' calls in the first 6 minutes. The owl is removed for a period varying in length from half a minute to 24 hours. Then it is put back for a second mobbing test and again the calls made in the first 6 minutes are counted. Obviously the shorter the gap between the two tests the less we expect the bird to respond on the second occasion. Fig. 1.5 shows that the strength of calling is up to 50% of its initial level if 30 minutes elapse between the tests, but a rest of 24 hours scarcely improves on this and Hinde finds that even longer rests never bring it back to its original level.

Now since an owl was the stimulus for both tests some of this lack of recovery could be due to stimulus-specific fatigue as with the begging of the young birds. However, Hinde found that using a stuffed stoat for the second mobbing test—a stoat is just as strong a stimulus as an owl—while it produced a small increase in calling, still left essentially the same picture as that shown in Fig. 1.5. Recovery from fatigue is rapid over the first 30 minutes, but thereafter it is extremely slow and is 'response-specific', i.e. it is some property of the mobbing response itself.

Fatigue of this type cannot result from the same mechanisms as does stimulus-specific fatigue. Hinde suggests that in fact there are two distinct processes here, one of which recovers rapidly whilst the other shows very little recovery. To these we must add the small degree of stimulus-specific fatigue demonstrated by exchanging owl for stoat. This means that perhaps three distinct fatigue and recovery processes are going on simultaneously when a chaffinch responds.

This is complexity enough, and although reflexes and complex behaviour both exhibit fatigue, the situation revealed by the chaffinch experiments must make us beware of jumping to conclusions about mechanisms. Each case must be studied separately to unravel the processes involved and only careful behaviour tests can do this.

Inhibition

This last category is not so much a property of reflexes as a whole aspect of the functioning of the nervous system. One of Sherrington's greatest contributions was the demonstration that sometimes impulses, passing along axons, inhibit transmission by the nerve cell with which they synapse, even though this cell may be simultaneously receiving excitatory impulses from other axons. Inhibition operates at every level within the central nervous system. Since every neuron is potentially in contact with every other, it is quite reasonable to consider why an animal is not thrown into a convulsion by every stimulus it gets. Part of the answer, as we have already seen, lies with the delay and fatigue that occur in synaptic transmission, but active inhibition is also involved.

Muscles are commonly arranged in antagonistic pairs, such that one flexes a portion of a limb and the other extends it. Sherrington showed that

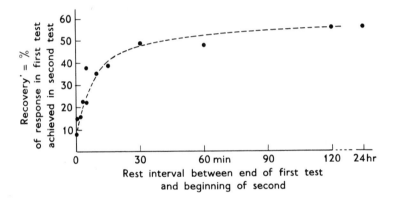

Fig. 1.5   The recovery of the owl-mobbing response after prolonged exposure to a stuffed owl; further explanation in text. (From Hinde,[108] *Proc. R. Soc., B,* **142,** 306.)

excitation of one member of a pair is accompanied by inhibition of its antagonist. Further, when the flexors of one limb are contracting, the flexors of the opposite limb are inhibited. *Reciprocal inhibition* of this type is one of the basic integrating mechanisms for walking.

With reflexes the active inhibition of a movement is easily distinguished from fatigue. The fatigue of the scratch reflex (Fig. 1.4) is marked by a gradual weakening. When the same reflex is inhibited by stimulating the antagonistic flexion reflex, it stops abruptly with no previous weakening and begins again the instant that the flexion stimulus is removed (Fig. 1.6).

Inhibition serves to bring about a smooth and rapid transition from one reflex action to another.

The rôle of inhibition in complex behaviour is superficially less obvious than that of excitation. We stimulate an animal and the most conspicuous result is that it makes a response. But in doing so it has made a swift transition which requires the inhibition of its behaviour prior to the stimulus until it adjusts its responses to the new situation. Sherrington considers the way in which reflexes appear to 'compete' for control of the final common pathway, i.e. that which controls the muscles whose action is common to several different reflexes. In an analogous way we may regard the different systems controlling patterns of complex behaviour like feeding, fighting and sleeping, as competing for control of the animal. Such patterns are obviously incompatible and only one can occur at a time. Which one does appear will depend on various factors both inside and outside the animal. When, say, the feeding system gains control, all others must be inhibited for the time being.

Sometimes we can identify parts of an animal's nervous system which have an inhibitory function, serving to suppress activity which would otherwise become incessant. Recent work on the brain of insects has revealed mechanisms of this kind. The brain selects appropriate responses by 'removing' inhibition from the relevant parts of the ventral nerve cord and its ganglia which then excite the required muscles. One of the most clear-cut examples of this comes from Roeder's[215] work with the praying mantis (Fig. 1.7). When a male mantis is decapitated, cutting off the brain from the ventral nerve cord, the body begins incessant stepping movements which carry it in a circle. In addition, the abdomen and genital appendages begin incessant copulatory movements and if it encounters a female, a headless male can often mate successfully. In nature males are sometimes decapitated as they approach the large, predatory females and these arrangements are highly adaptive!

In vertebrates, too, we often find that the result of destroying some area of the brain is to make one type of behaviour much easier to evoke. As we shall see in later chapters, some caution is needed in interpreting the results of brain damage experiments, but nevertheless there can be no doubt that parts of the vertebrate brain regulate the activity of others by inhibition.

Sherrington found that when inhibition was removed from a reflex it returned at a higher intensity than it had previously. Fig. 1.6 shows this phenomenon, which Sherrington called 'reflex rebound', for the scratch reflex. We commonly observe that when a particular type of complex behaviour—for example courtship—has not been elicited for some time, it has a lowered threshold and is performed with high intensity when it is, at last, evoked. It is possible that the system controlling courtship has been inhibited by those of other activities and shows something akin to reflex rebound when this inhibition has been removed. Kennedy[136-7] has interpreted some aspects of the behaviour of aphids along these lines. The

A

Scratch reflex stimulus

B

Flexion reflex stimulus

Seconds

Fig. 1.6 Inhibition of the scratch reflex by the flexion reflex. The stimulus denoted on line A evokes the scratch reflex, but this response is inhibited when the stimulus on line B evokes the flexion reflex. The moment B is removed the scratch reflex returns, and much more vigorously than before—an instance of 'reflex-rebound'. (Modified from Sherrington,[241] 1906, *The Integrative Action of the Nervous System*. Charles Scribner's Sons, New York.)

behaviour of winged aphids alternates between periods of flight and periods of settling and feeding upon leaves. If an aphid settles on an 'unattractive' surface—an old leaf, for example—it does not stay long and soon takes off but flies relatively weakly and soon settles again. Conversely, if it has settled on an attractive young shoot it stays for a long period but, when it takes off, flies vigorously and for a long time.

By an elegant series of experiments, Kennedy[137] has been able to exclude any simple explanation for this relationship based on physical exhaustion during flight and recovery after resting and feeding on a young leaf. He suggests that there is mutual inhibition between the systems controlling flight behaviour and those controlling settling. As with the reflexes, activation of the settling system may temporarily inhibit the expression of the flight system but, at the same time, gradually lower the threshold for flight.

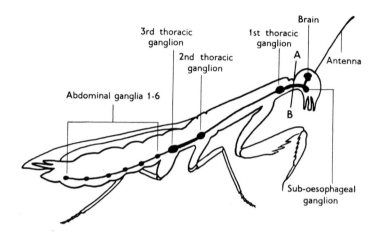

Fig. 1.7   Diagram of the central nervous system of the praying mantis. In Roeder's experiments the nerve cord was cut along the line AB. (From Roeder,[215] 1963, *Nerve Cells and Insect Behavior*. By permission from Harvard Univ. Press, Cambridge, Mass. © by the President and Fellows of Harvard College.)

Kennedy calls this relationship 'antagonistic induction' and it clearly resembles reflex rebound in some ways.

In Chapters 4 and 5 we shall discuss other evidence that the systems controlling complex behaviour do inhibit one another and that particular patterns 'break through' when their inhibition is weakened. However, reflex rebound cannot provide the whole answer to the problem of how one type of complex behaviour comes to 'take over' from another. Other factors can cause thresholds to be lowered. For example, it is quite inappropriate to speak, in physiological terms, of the 'inhibition of drinking' if a dog is physically prevented from reaching its water bowl, but the dog nevertheless becomes increasingly responsive to water.

This chapter has tried to show that reflexes and complex behaviour share many properties. Of course one cannot always explain behavioural observations in reflex terminology and the attempt is often pointless. There *is* a difference in complexity and for the present one has usually to be content with an analysis in terms of behaviour, leaving most of the physiology to come later.

Even if much of the physiological background remains vague, in what follows we must always bear in mind the constant activity of the nervous system. It does not lie dormant requiring a stimulus to provoke it into a response. The sense organs and the proprioceptors in the muscles provide a continuous background flow of impulses whose level is affected both by external stimuli and by efferent impulses from the brain. Information from the external world and the muscles is combined by the central nervous system with information on various aspects of the body's metabolic state— temperature, carbon dioxide concentration and water balance, among others. Any shift from the optimum is rapidly corrected, and most of the overt behaviour we observe as a result of all this neural activity is adaptive to the animal's survival and designed to restore the *status quo*. Behaviour is one of the most important homeostatic mechanisms, one that is both sensitive and powerful.

2

The Development of Behaviour

Nearly all the behaviour we observe in animals is adaptive. They respond to appropriate stimuli in an effective manner and thereby feed themselves, find shelter, mate and rear families. Animals are certainly not infallible, but when they do make mistakes it is often because they have been transported into an unnatural environment. We are not surprised if birds make futile attempts to escape when first put into a cage.

INSTINCT AND LEARNING

How does an animal's behaviour become so well fitted to its normal environment? There are two basic ways. Firstly, it may be born with the right responses 'built in' to the nervous system as part of its inherited structure. Honey-bees inherit the ability to form wings and wing muscles for flight; they also inherit the tendency to fly towards flowers and seek nectar and pollen. Such responses are popularly called 'instinctive'—a term which has often been abused but remains useful. Instinctive behaviour evolves gradually, as do structural features, and natural selection modifies it to fit the environment in the best way. It forms a kind of 'species memory' passed on from each generation to its offspring.

Alternatively an animal may be born with few inherited responses but, instead with an ability to modify its behaviour in the light of its experience as it grows up. It learns which responses give the best results and changes its behaviour accordingly.

Instinct and learning both ensure adaptive behaviour, the former by selection operating during the history of a species, the latter during the history of an individual. Stated in this way there is a clear dichotomy, which is rather unrealistic when actual examples are examined. But before we do this it is worth considering the importance of instinct and learning in a general way through the animal kingdom.

Instinct and learning in their biological setting

Instinct can equip an animal with a series of adaptive responses which appear ready-made at their first performance. This is clearly advantageous for animals with short lifespans and little or no parental care. The arthropods, for example, show a remarkable development of instinct for no other course is open to them. A female digger-wasp emerges from her underground pupa in spring. Her parents died the previous summer. She has to mate with a male wasp and then perform a whole series of complex patterns connected with digging out a nest hole, constructing cells within it, hunting and killing prey such as caterpillars, provisioning the cells with the prey, laying eggs and finally sealing up the cells. All this must be completed within a few weeks, after which the wasp dies. It is quite inconceivable that she could achieve this tight schedule if she had to learn everything from scratch and by trial and error.

Contrast the digger-wasp's situation with that of a lion cub. Born quite helpless, it is sheltered and fed by its mother until it can move around. It is gradually introduced to solid food and gains agility in playing with its litter mates. It has constant opportunities to watch and copy its parents and other members of the group as they stalk and capture prey. It may catch its first small live prey when 6 months old, but it is 2 years or more before it is fully grown. Its behaviour, and particularly the methods and stratagems it uses in hunting, may change according to circumstances throughout its life.

The digger-wasp which must rely on pre-set instinctive behaviour and the lion who can learn in relative leisure represent two extremes on the behavioural scale. In fact the preceding descriptions have greatly oversimplified their actual behavioural development. The digger-wasp can and must learn many things during its brief life—the exact locality of each of its nests, for example, so that it can return to them after its hunting trips. The young lion possesses some predatory tendencies which are certainly instinctive, even though it has to learn how to direct them.

All animals above the annelid worm level show both kinds of behaviour and each has its own special advantages. This is clearly illustrated from studies on bird calls. Birds often show a strong development of both instinct and learning ability. As we shall discuss later, the song of a male bird often requires the experience both of singing and listening to other males before it takes on its final form. By contrast, in every bird species studied both the production of and response to the alarm calls of the species appear perfect at the first showing. Natural selection has favoured an inherited response where the delay of learning may prove fatal.

The chief advantage of learning over instinct is its greater potential for changing behaviour to meet changing circumstances. Such a consideration is obviously more important to a long-lived animal than to an insect which lives only a few weeks. A further relevant factor may be body size, because highly-developed learning ability requires a relatively large amount of brain tissue, insupportable in a very small animal. Usually body size and life span are positively correlated to some extent and large animals live longer than small ones. Apart from these physical considerations it is clear that natural selection can produce different degrees of learning ability to match a species' life history. The two most advanced orders of the insects, the Hymenoptera (ants, bees and wasps) and the Diptera (two-winged flies), are comparable in size and life span. The Hymenoptera, in addition to a rich instinctive behaviour repertoire, show an extraordinary facility for learning, albeit of a simple type, and this plays an important rôle in their lives. During her brief 3 weeks of foraging a worker honey-bee will learn the precise location of her hive and the locations of the series of flower crops on which she feeds. She may move from one to another of these during the course of a day's foraging because she also learns at what time of day each is secreting the most nectar.

On the other hand, in spite of many attempts, there has been scarcely any satisfactory demonstration of learning from the Diptera, although they do show certain types of habituation (see p. 177). Their short lives are governed with complete success, by inherited responses to food, shelter and a mate.

The characteristics of instinct and learning

We must now examine more closely the characteristics of instinct and learning. There are two conspicuous features of instinctive behaviour which may seem, at first sight, to be unique. First, that it consists of rigid, stereo-typed patterns of movement which are very similar in all individuals of a species; all digger-wasps of the same species build their nests in the same way, domestic cockerels all use the same series of movements when courting hens, and so on. Secondly, instinctive patterns can often be evoked most readily by very simple stimuli. When presented with a complex situation the animal responds to one part of it and virtually ignores the rest. A robin displays more aggression to a tuft of red feathers from the breast of a rival male than to a complete bird which lacks only these feathers.

However striking, such characteristics are quite inadequate to distinguish instinctive from learnt behaviour. The latter is often described rather vaguely as being 'more flexible', but in fact the patterns of movement involved may be just as stereotyped as those of instinct. Rats placed in a box where they must learn to press on a projecting lever in order to obtain a pellet of food (Fig. 2.1) will develop a particular manner of doing so. Some use always their left paw, others press with their chins, and individual rats are very conservative in the method they use. In his delightful book, *King Solomon's Ring*, Lorenz[158] describes how water shrews learn the

Fig. 2.1   Rat in a Skinner box pressing the lever which delivers a small pellet of food into the cup.

geography of their environment in amazing detail. If at one point on a trail they have to jump over a small log, this movement is learnt with such fixity that they continue to make the jump in precisely the same fashion long after the obstacle is removed. There are plenty of other examples where animals have comparable difficulty in 'unlearning' something. The movements appear to become almost 'automatic' and may persist even if they are no longer effective. Learnt patterns nearly always involve responding to particular cues in the environment and, just as with instinctive behaviour, other features may be ignored. It is relatively easy to train animals to discriminate one key stimulus within a complex changing situation.

If we are to make any critical distinction between instinctive and learnt behaviour, it must be based not upon their overt characteristics but upon their development within the individual. Rats learn to press a food lever and each does so in a stereotyped way, but this way is unique to the rat and varies between rats. All sexually receptive female rats show the same stereo-

Fig. 2.2 Copulation in rats. The female's posture with raised pelvis and deflected tail is very stereotyped in form. The pressure of the male's fore-limbs on her flanks is one of the stimuli necessary for the female to respond. (From Barnett,[14] 1963, *A Study in Behaviour*. Methuen, London.)

typed acceptance posture when mounted by a male (Fig. 2.2). Furthermore if we put a hungry rat into a box with a food lever, it may be hours before, by chance, it pushes against the lever. But if a virgin female rat is brought into receptive condition by a hormone injection, it assumes the acceptance posture on the first occasion it is put with a male.

Here we have the basis for developmental criteria of instinctive behaviour and the commonest way of testing them has been the so-called 'isolation experiment'. Animals are kept individually out of contact with others from as early an age as possible. When mature, their responses to a variety of stimuli are tested and compared with those of animals reared normally. Rather few animals have been tested under really rigorous conditions but some fish and birds have been shown to perform various feeding, sexual and alarm patterns of behaviour quite normally after being reared in isolation. The life histories of many insects are natural isolation experiments which demonstrate vividly that no practice or learning are involved in the behavioural development of the adults. However easily isolation experiments enable us to eliminate the possibility that animals learn how to do something, they offer only a very restricted view of the factors that may be operating during development—learning and practice are only two from a wide range of possibilities.

Behaviour patterns that develop normally in complete isolation must be under some degree of genetic control and result from an inherited potentiality of the animal's nervous system (see further discussion in Chapter 7). But if we leave our consideration of the development of instinct at this point we are losing a great deal of important information. Genes may control behaviour, but they must interact with developing animal's environment. It is obvious that an animal carrying genes for high body weight will not develop its potential if it is half-starved from birth. We may discover factors in the behavioural environment of an animal which influence the development of instinct. Clearly we must avoid any simple dichotomy which ascribes instinct to the genes and learning to the environment. Both must be involved in the development of all behaviour. This point might not seem to need such emphasis, but in fact misunderstandings here have been, until quite recently, the basis for a considerable dispute among animal behaviour workers.

Ethologists and psychologists

After World War II there were two main schools of animal behaviour in the West. The first consisted largely of American experimental psychologists who could claim descent from J. B. Watson,[275] whose book *Behaviorism*, published in 1924, was a landmark in the modern experimental approach to behaviour. The second school—largely European—was founded in the late 1930's by Konrad Lorenz, an Austrian working in Germany. He was joined by Niko Tinbergen from the Netherlands, who collaborated with him in some early work. When the War ended both men rapidly built up groups of research workers and others, usually zoologists, became associated with their approach. This group call themselves 'ethologists', and ethology has been defined simply as 'the scientific study of behaviour'.

Ethologists and American psychologists approached their subject from quite opposite ends. Ethologists studied a wide range of animals under near natural conditions, often in the field with the use of hides. They were impressed by, and concentrated on, the rich variety of adaptive behaviour patterns shown by insects, fish and birds. Reproductive behaviour which is often conspicuous and easy to observe formed a large part of their studies, and ethologists often have great success in keeping animals and getting them to breed in captivity.

The original ethological approach is set out by Tinbergen[258] in his book, *The Study of Instinct*. It is, in essence, the experimental elucidation of the stimuli which evoke and the motivation which controls the performance of instinctive behaviour, together with a study of its survival value and its evolution. This last aspect is studied in some detail by comparing the behaviour of closely related species and deducing the evolution of particular behaviour patterns, just as a comparative anatomist does for structural features. Ethologists were not much interested in problems related to learning and, though they recognized that their animals could learn, they generally ignored or eliminated this fact from their theories and experiments.

Experimental psychologists, on the other hand, were almost exclusively interested in learning. They largely ignored an animal's behaviour in its natural environment and deliberately confined their subjects to experimental conditions. A maze; a 'shuttle-box' in which it learns to move from one side to another to avoid electric shock; a 'Skinner box' in which it learns to press a lever or peck at a key to receive a reward of food or drink; these were the situations that interested the psychologists. Their aim was to construct 'laws of behaviour' which would describe how an animal's behaviour changes after given levels of practice, reward or punishment and to make predictions, for example, about the efficacy of one or another experimental situation for producing learning.

It is not surprising that the experimental psychologists used a narrower range of animals than the ethologists. Mammals are the best animals for their type of learning study, and were used almost exclusively, with domesticated albino rats providing the great majority of the subjects.

At first the two schools took little note of each other's findings, but after some years cross-fire began and each tended to adopt rather extreme positions. See, for example, Lehrman[147] for an extreme statement of the case against the ethological concepts then current. The psychologists accused the ethologists, among other things, of grossly underestimating the rôle of the environment in the development of behaviour, of regarding the label 'instinctive' as an adequate explanation in itself and of applying this label far too readily and without sufficient evidence. Conversely, ethologists complained that most experimental psychologists were ignorant of the behaviour of any animal other than the white rat. They suggested that the reality of instinct would become obvious to anybody who could drag himself away from a Skinner box to watch a colony of honey-bees or a nesting male stickleback.

Both sides scored good points and both have subsequently benefited from a dispute which for once did generate more light than heat. Ethologists are now much more cautious in their use of terms and increasingly turn their attention to the development of behaviour. Experimental psychologists are employing a wider range of animals and studying them in a wider range of situations. In particular the studies of physiological psychologists, who mostly remained outside the dispute, now form an excellent link between the two approaches.

THE DEVELOPMENT OF BEHAVIOUR WITHIN THE INDIVIDUAL

The majority of workers would now agree that any rigid classification of behaviour into 'instinctive' or 'learnt' is inadequate, but that all behaviour presents us with problems of development. What processes are involved in the emergence of fully-formed adult behaviour? This is a very wide topic and here we can only take a few examples to illustrate some of the complex range of influences that must be considered. They will serve to emphasize how meaningless it is to try to press behaviour into categories.

Some general aspects of early experience

Before going on to deal with the development of specific behaviour patterns, it is important to recognize some general features of behavioural development (see also the discussion on p. 202). Young animals are extremely vulnerable to changes in their environment the effects of which can often be detected later in life. For the embryologist development begins when the egg is fertilized, but the majority of behavioural experiments begin much later with animals which, though they may be young, are relatively independent. Consequently we may overlook the earliest stages of development such as the young bird goes through in the egg or the young mammal inside its mother. It is certainly not immune from factors that affect its behaviour even then. Gottlieb[85A] has carefully followed the development of behaviour in bird embryos and finds, for example, that a young duckling can respond to sounds played to it at least a day before hatching. Its heartbeat accelerates and it may call more itself. Further, its response to the calls of its mother is reduced if it has been isolated during this pre-hatching period and thus was unable to hear the calls made by other embryos inside their shells. Vince[269A] has shown that quail embryos respond to the clicking calls made by the other members of the clutch. Remarkably, this communication between embryos serves to synchronize their hatching. Slow developers are accelerated by hearing the calls characteristic of advanced embryos and, to a lesser extent, advanced embryos are slowed down by the calls from less advanced neighbours.

In mammals the embryo is more insulated from the external world but, of course, more directly dependent on its mother's physiological state. If female rats are kept under stress during pregnancy the behaviour of their offspring is affected.[131C,252D] The type of change produced concerns the degree of fearfulness that they show when first put into a strange environment. Similar effects can be produced by a whole variety of treatments to young animals after birth. Mild electric shock, cooling or even the simple act of lifting them from the nest and then replacing them can affect their subsequent behaviour as adults. Rats which have been handled during infancy mature more quickly—their eyes open sooner and they are heavier at a given age than unhandled controls. In addition they appear less 'emotional' in frightening situations. By this we mean, for example, that when placed in a large, brightly lit arena they move around and explore more than unhandled rats, which spend more time crouching and cling to the edges of the arena if they do move. The way in which early handling produces such effects in adults is not at all clear and is currently an active field of research (see Levine et al.[151]). The 'extra' stimulation of young animals may act on them directly so as to change their subsequent development, but more probably it also affects them through their mother. She reacts to their changed condition after handling or shock by giving them more attention than untreated young. Barnett and Burn[16] have some direct evidence of this with baby rats that had been marked by having their ears punched. Just after they were returned to the nest such babies received three times as much nuzzling and licking from their mothers as did untouched babies.

In mammals and some birds it is clear that a most subtle interplay between the mother and her offspring affects the way their behaviour develops. We must expect that interfering with this relationship will have long lasting and perhaps permanent effects. There is now evidence that the effects of handling rats in infancy can be transmitted at least as far as their grandchildren.[61C] Presumably handled rats have altered maternal physiology and behaviour which affects the early experience of their offspring whose maternal behaviour is, in turn, affected with results still detectable in their own young.

If we know little of the underlying mechanisms in such cases, there are others where it is possible to identify the basis of behavioural development more clearly.

Development involving growth or maturation

The development of an animal's behaviour must obviously be linked to its normal growth processes. For example, the development of sexual behaviour in most vertebrates is linked with the growth of the gonads. Sometimes improvement in the performance of a behaviour pattern can be associated with the development of the animal's nervous system and such improvement is usually called 'maturation'.

The behaviour of embryos is determined by their stage in development and their increasing complexity of structure as they develop is paralleled by an increasing repertoire of behaviour, both spontaneous and in response to external stimuli. One of the classic examples is Carmichael's[43-4] study on developing frog tadpoles. Whilst still in the jelly they begin flexing their tails back and forth as in swimming and these movements get more vigorous and complete with time. Is this improvement due to practice or simply to growth? Carmichael kept some developing tadpoles under continuous light anaesthesia which did not slow their growth but prevented all movement. When brought round and released at the age of emerging from the jelly they swam just as proficiently as unanaesthetized tadpoles, which implies that maturation and not practice is the cause of the changes seen in the behaviour of normal tadpoles.

Similar experiments have been made with birds which often appear to 'practice' flying by beating their wings whilst still in the nest. In 1873 Spalding[249] showed that young swallows reared in cages so small that they could not stretch their wings flew just as well as normally reared birds when released. The ability to fly can be said to 'mature' without practice in these cases, but young fledglings cannot fly as well as adult birds and all the finer points of flying skill are added by practice.

In other cases practice is required from the outset if behaviour is to develop properly. In Chapter 1 Wells'[279-80] experiments with young cuttle-fish were mentioned. It will be recalled that their response to a tiny shrimp presented in a glass tube can be divided into four stages (see Fig. 2.3):

1. A latency before there is any observable response (Fig. 2.3a).
2. The nearest eye of the cuttle-fish fixates the shrimp.
3. The cuttle-fish turns towards the shrimp and both eyes fixate (Fig. 2.3b).
4. It attacks and seizes the shrimp (Fig. 2.3c).

Stages 2, 3 and 4 usually occupy about 10 seconds and this time varies very little with age or experience. The duration of stage 1 shows a rapid decline with successive tests as shown in Fig. 2.4. After five trials at a rate of one per day, the latency is reduced from about 120 seconds to 10 seconds or less. This change is the same whether the attacks are successful or unsuccessful, whether made with a 1-day-old cuttle-fish or one starved for 5 days before its first test. The only common factor appears to be practice in attacking shrimps. Learning is ruled out because in some other tests Wells finds it impossible to show any sign of learning ability until the cuttle-fish are a month old. Baby cuttle-fish go on battering away at a glass plate behind which they can see a shrimp, until they are physically exhausted.

Nor are they deterred from such attacks by electric shocks. With similar treatment an adult will stop attacking after a couple of trials. The vertical lobe of the cuttle-fish brain, which is known to be concerned with learning in adults (see p. 225), does not develop until relatively late after hatching.

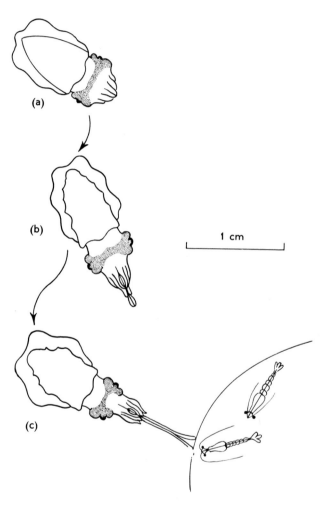

Fig. 2.3 The attack of *Sepia* on *Mysis* : (a) Newly-hatched *Sepia* at rest ; (b) swimming towards the prey, the eyes turned forward and fixating it ; (c) stabbing with the long pair of tentacles at the *Mysis* behind glass. (From Wells,[279] 1958, *Behaviour*, **13**, 96.)

The development of pecking in newly-hatched chicks provides another example of the interaction between maturation and practice. Young chicks have an inherited tendency to peck at objects which contrast with the background, but their aim is at first rather poor, and various workers have studied how it improves. One of the most complete studies was made by Cruze.[56] He hand-fed chicks in the dark on powdered food for periods of

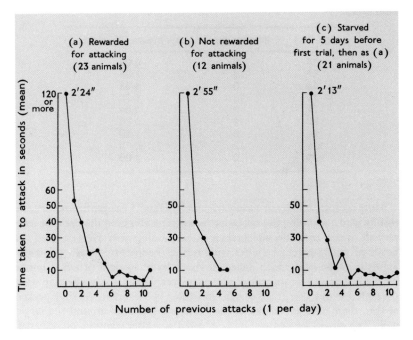

(a) Rewarded for attacking (23 animals) (b) Not rewarded for attacking (12 animals) (c) Starved for 5 days before first trial, then as (a) (21 animals)

Time taken to attack in seconds (mean)

Number of previous attacks (1 per day)

Fig. 2.4 The rapid decline of latency to attack *Mysis* in newly-hatched *Sepia*. The latency depends on the number of attacks that have already been made and is not affected by reward (a & b) food deprivation or age (c). (From Wells,[279] 1958, *Behaviour*, 13, 96.)

up to 5 days before testing the accuracy of their pecking. Whilst in the dark they are inactive and have no chance to practise the movement. Cruze measured accuracy by putting the chicks individually into a small arena with a black floor, on to which he scattered two or three grains of millet. Each chick was allowed 25 pecks scored for miss or hit and grains were replaced if the chick swallowed them. After accuracy tests the chicks were allowed to feed naturally in the light and the effects of practice on their accuracy measured again after 12 hours.

Table 2.1 shows the results of one experiment. There is a steady improvement with age (heavy type), but at any age 12 hours of practice greatly improves accuracy (light type).

Table 2.1   The pecking accuracy of chicks at different ages before and after 12 hours of practice. Each figure represents an average from 25 chicks. (Modified from Cruze.[56])

Age (hr.)	Practice (hr.)	Av. misses (25 pecks)
24	0	**6·04**
48	12	1·96
48	0	**4·32**
72	12	1·76
72	0	**3·00**
96	12	0·76
96	0	**1·88**
120	12	0·16
120	0	**1·00**

More recently an ingenious experiment by Hess[104] has confirmed Cruze's results whilst avoiding the necessity for keeping chicks in the dark for long periods, which probably affects their visual development. Hess covered the heads of newly-hatched chicks with rubber hoods fitted with prismatic lenses over the eyes which deflected vision by 7° to right or left. Control chicks had plain transparent plastic eyepieces. When 1 day old they were presented with a bright silver nail embedded in Plasticine. As they pecked at this, their bills left marks in the Plasticine scattered around the target nail. The marks of control birds showed a scatter centred on the nail itself; those of birds with deflected vision showed a similar degree of scatter but centred around a point to the left or right, according to which side the prisms deflected (Fig. 2.5a and b). For the next 3 days all the chicks fed themselves from bowls of mash or scattered grain. Their pecking patterns at 4 days old are shown in Fig. 2.5c and d. Both groups show a similar improvement in accuracy, but the prism group had not learnt to correct for the displacement, indeed many of them had lost weight because they couldn't feed themselves well. Their improvement could not have been due to practice, because they never hit the grain at which they were aiming but, if any, another near by. Maturation thus seems the only explanation, although we must bear in mind that mechanisms other than those specifically controlling pecking will influence accuracy. The chicks' legs grow stronger and perhaps their stability improves which would help their aim.

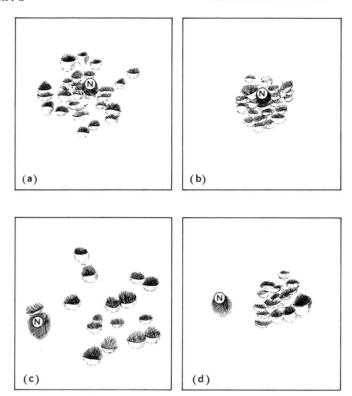

Fig. 2.5 The accuracy of pecking in chicks as revealed by the marks of their bills in Plasticine around a 'target' brass nail head (N). Each picture is the record of a single chick. (a & b) are from a normal chick at 1 and 4 days old respectively; its accuracy improves greatly. The chick in (c) and (d) wore a hood which deflected its vision 7° to the right. Although it never hits the target its accuracy improves with age in just the same way as in normal chicks. (After photographs from Hess,[104] 1956, *Scientific American*, **195** (1), 71.)

Fixed action patterns

When discussing the results of handling and other forms of mild stress during infancy, we could identify effects which we may regard as 'non-specific', i.e. they affect the animal's responsiveness in a number of different situations, although we may try to ascribe this to some more basic effect such as a change in emotionality. However it is common to find that certain experiences have a very specific effect on subsequent behaviour. Thus stimuli which a young animal perceives during the process of sexual imprinting (see p. 198) affect its choice of a sexual partner when it matures.

If *Drosophila* larvae are reared in a food medium containing geraniol, the adult flies that emerge have their normal aversive reaction to its scent changed to one of indifference, although their response to other scents is not affected.[178A]

In general it seems easier for early experience to modify the stimuli to which an animal responds in later life than to modify the form of the behaviour patterns it employs. Cockerels raised under hens or in incubators, isolated in battery cages or kept in flocks may differ markedly in the range of objects they will court and the frequency with which they do so, but they all show the same stereotyped movements immediately recognizable as 'waltzing', 'tidbitting' and 'mounting' by those who study the behaviour of the fowl.[285]

Such movements, which are shared by all members of a species, are commonly referred to as 'fixed action patterns'. We know next to nothing of how they develop, because they usually appear in their typical form at the very first performance. Having said this, it must be admitted that we have very few detailed studies which either compare the exact form of a fixed action pattern in different members of the same species, or examine the way it is performed at different stages through the life of an individual. This is a laborious task involving the analysis of slow-motion film or videotape. Where we have such records for gulls and ducks and *Drosophila* they do reveal some variation between individuals and some changes in the exact form of a movement as an animal grows up—often one supposes because its muscles and skeleton are growing too. However, the most striking feature of Dane and van der Kloot's[60A] study on courtship movements of the goldeneye drake is the astonishing constancy of each fixed action pattern within an individual. For example, a pattern called the 'simple head throw' had an average duration of 1·29 seconds and never deviated more than a few *hundredths* of a second from this.

Occasionally we can observe young animals perform recognizable fragments of fixed action patterns long before they become integrated into the full adult pattern. Nestling cormorants, holding nothing in their beaks, make the head and neck movements which form part of the 'tremble-shove' pattern used by the adult during nest-building.

Because of the spontaneity and lack of obvious practice in such cases, ethologists have tended to regard fixed action patterns as being the closest thing to pure inherited behaviour. Certainly there must be an important inherited component in the development of such patterns and we shall discuss some genetic evidence for this in Chapter 7. To some extent the fixed action patterns are determined along with the developing structure of the animal. The precise and very constant form of the courtship wing movements of *Drosophila* males is largely determined by the structure of their thorax and its musculature.[71B] But in addition to structural considerations, the developing nervous system must lay down circuits which predispose the animal to perform this or that particular sequence of muscle contractions that go to make up a fixed action pattern. Even so we can sometimes identify environmental factors which are vital for such development, and they are not always obvious ones.

Many reflexes—which are as good examples of fixed action patterns as can be found—do not develop normally in restricted environments. Riesen[213] reared chimpanzees from birth in darkness for 40 months. When first tested in the light they showed no eye-blink reflex and this took some 5 days to develop. Clearly the presence of light is necessary for the development of this inherited potentiality. Reyniers[211] reports another interesting case which was discovered by accident. Infant rats can be removed from a pregnant female by Caesarean section and transfered to a completely sterile environment for studies of germ-free life. When this was first attempted all the babies died after a day although they began to feed well. Post-mortem examination showed that their bladders had ruptured. It turned out that some initial tactile stimuli to the genital region is necessary to start the reflex emptying of the bladder—normally provided by the mother rat. Once set going by this 'trigger' from the environment, the reflex subsequently operates on its own when the bladder is full.

In the examples given above, conventional learning has been eliminated as contributing to the development of inherited behaviour. However, it is very common to find that adult behaviour patterns contain elements of both. Lorenz[159] and, following him, Eibl-Eibesfeldt[69A] continue to emphasize the importance of attempting to distinguish between inherited and learnt aspects of behaviour. Lorenz suggests that learning can only affect development in a limited way and that, in particular, it cannot modify the fixed action patterns themselves. Certainly an animal's ability to learn novel patterns of movement—new motor skills as we might call them—is circumscribed by its inherited behaviour. Many ground-nesting birds retrieve an egg which has slipped out of the nest in the same way. They extend the neck and hook the lower mandible over the egg which is then rolled back by drawing in the head (see Fig. 2.6). This fixed action pattern can be elicited on the first occasion that an incubating bird is presented with an egg outside the rim of the nest. For a goose to use the bill in this way is reasonably efficient because it has a broad one, but birds with narrower bills such as gulls and waders often have considerable trouble because the egg tends to slip out to one side. It would seem much easier for these birds to use their wing or foot as a scoop, but they show no variation from the fixed pattern.

Perhaps the most vivid example of how resistant inherited fixed action patterns are to change, comes from the work of Dilger[67] with love-birds. These are members of the parrot family which breed readily in captivity.

Within the genus *Agapornis* two types of nest-building behaviour are represented. All species tear strips of material from leaves to build with (in the laboratory newspaper forms an excellent substitute) but whilst some tuck the strips into their rump feathers and fly back to the nest carrying several pieces at once, other species carry the strips singly in their bill. Dilger crossed two species which differed in this respect and watched the nest-building behaviour of the hybrids. For some time such birds were incapable of building a nest at all because they attempted to perform some kind of compromise between the two collecting methods. They might start

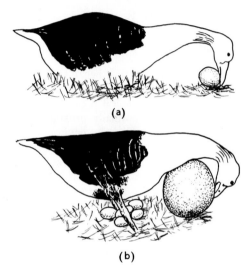

(a)

(b)

Fig. 2.6 Grey lag goose retrieving an egg which is outside the nest. This movement is very stereotyped in form and used by many ground-nesting birds. The goose attempts to retrieve a giant egg in precisely the same fashion. (From Lorenz and Tinbergen,[161] 1938, *Z. Tierpsychol.*, **2**, 1.)

to tuck a strip between the rump feathers, but either failed to let go of it or failed to tuck it properly. The end result was usually that the strip fell to the ground and the whole process began again. The only success the hybrids had was when they managed to retain a strip in their bill after the attempted tucking procedure. Dilger found that even after months of practice the birds managed successful carrying in only 41% of their trials. Two years later they were successful in nearly all trials, but before carrying a strip in their bill they still made the turning movement of the head which is the preliminary to tucking. Parrots are intelligent birds which are known to learn quickly in other situations, but here the inherited predisposition to perform the tucking sequence outweighed learning for a long time. More recently Buckley[40] has repeated and extended Dilger's observations, and he describes a number of other ways in which the behaviour of the hybrid birds is disrupted.

Now we must turn to consider some examples of behavioural development in which learning and inherited tendencies interact more fluently.

Development which involves the interaction of inherited tendencies with learning

We have previously mentioned the manner in which the basic pattern of bird flight matures in the absence of any practice whereas the finer skills of flying develop later. By close observation of young animals and controlling the time at which they are first exposed to otherwise natural situations, it is sometimes possible to identify analogous stages in their behavioural development.

Eibl-Eibesfeldt[69A] describes a number of examples from his studies of young mammals. He reared young polecats in isolation and never gave them any opportunity to catch live prey. Such animals showed varying degrees of interest when first presented with a live rat, but no hint of attack unless the rat ran away from them. If it did so, they instantly pursued and seized it in their mouth, usually shaking it rapidly in a characteristic way. Their bites were at first badly orientated, but after a few trials they were seizing prey by the nape of the neck and killing with a single bite. Clearly there are inherited components to the killing pattern which are completed by learning. Eibl-Eibesfeldt found that normally polecats picked up the necessary practice during play sessions with their litter mates. In a similar series of observations with hand-reared squirrels, he has shown that although they respond to nuts and try to open them on the first exposure, their efforts are uncoordinated. They have to learn to direct their gnawing to the thinnest portion of the shell and to confine it to this one area.

One of the most beautiful examples of the dovetailing of inherited and learnt components during development is that of bird song. Until recently song was difficult to study because there was no way of expressing it graphically. The development of the sound spectrograph has turned this situation inside out and made detailed analysis easier for song than for most behaviour. Bird song is just like any other behaviour pattern—a controlled sequence of muscle contractions which in this case we perceive as sound. It can be recorded on tape and played into the spectrograph which produces a chart showing how much energy was emitted at the various sound frequencies at any time.

Marler and Tamura[180] have worked on the song development of the American white-crowned sparrow. This small finch has a wide range on the Pacific coast, and birds from different regions have recognizably different song 'dialects'. If young males are taken immediately after hatching and reared alone in sound-proof chambers then, no matter which region they come from, they all eventually sing very similar and simplified versions of the normal song. Obviously they must pick up the local dialect by listening to adult birds and modifying their own simple song pattern accordingly. Marler and Tamura found that this learning process usually takes place during the first 3 months of life and thus before the bird has ever sung itself. Males captured in their first autumn and reared alone begin to sing for the first time with a recognizable version of the local dialect. Up to 3 months of age, isolated males can be 'trained' to sing their own or other dialects by playing to them tape-recorded songs, though the results of such training do not show until the birds begin to sing themselves some months later. Beyond 4 months of age the birds are unreceptive to any further training and their songs when they begin singing are not affected. Here then we have a simple, inherited song pattern which is sensitive to modification by learning but only during early life. The young birds 'carry' the memory of the songs they hear and reproduce them when they first sing.

The analysis has been taken further by the experiments of Konishi.[141] He completely deafened birds at various ages by removing the cochlea of the inner ear. If this is done to a young fledgling just out of the nest, it will subsequently sing, but it produces only a series of disconnected notes. These are quite unlike the song of isolated birds which, though simple in form, would still be recognizable as 'white-crowned sparrow' to an ornithologist. The bird has to be able to hear itself in order to produce this inherited song pattern. In other words, it would be more accurate to say that it is not the capacity to produce the simple song which is inherited, but rather some kind of neural template representing this song, against which the bird matches the notes which it produces and adjusts them to fit. If Konishi deafened young birds *after* they had been 'trained' with normal song but *before* they had themselves sung, their subsequent song resembled that of birds deafened as fledglings. They need to hear themselves in order to match up the song they produce with that which they have stored in their memories. Presumably the songs they have heard as fledglings have modified the inherited template so that it conforms to the more complex characteristics of normal adult song. Once the birds have matched their own output with this and sung the adult song, they can go on singing normally even when deafened. At this stage song development comes to an end in the white-crowned sparrow; after its first spring the bird is no longer susceptible to further experience and keeps much the same song pattern for the rest of its life. However, one surprising feature of bird song development is how much it varies between species, so that it is hard to generalize. The song of chaffinches and zebra finches develops in much the same way as the white-crowned sparrow, but Oregon juncos[179C] and indigo buntings[212A] continue to modify their songs according to experience for at least another year, and blackbirds may change the details of their songs throughout their life[94] (see also reviews in Hinde[112A]).

The results of the experiments with the white-crowned sparrow are summarized in Fig. 2.7. We have gone over them in some detail because they are an object lesson in the study of behavioural development. They

Fig. 2.7 A summary of the results obtained by Marler, Tamura and Konishi on song development in the white-crowned sparrow.

show how subtle and complex are the interactions between inherited tendencies and the environment which finally produce behaviour as we observe it.

Fig. 2.7 also indicates the result of one more experiment with the sparrows which introduces the last aspect of behavioural development to be considered.

The inheritance of a predisposition to learn particular things

White-crowned sparrows only learn to sing the white-crowned sparrow song. If the songs of other species are played to males during their 'sensitive period' for learning they have no effect and the birds' subsequent songs sound like those of isolated males. The chaffinch behaves very similarly, and Thorpe[254] points out that this can scarcely be a result of limitations in its sound-producing organ or syrinx. The related bullfinch and greenfinch have syrinxes of almost identical structure. They have only a poorly developed natural song but are good mimics and will learn to reproduce the songs of many other birds.

Some clue to the chaffinch's limitation is provided by the song of the one species it will learn to imitate reasonably well. This is the tree-pipit, whose song to the human ear also has a chaffinch-like tone, though it is very different in pattern. The chaffinch must have an inherited tendency to single out this tone from all others and to reproduce it.

There are several other cases on record where animals show a singular facility for particular types of learning. Tinbergen's[260] work with the herring-gull has shown that whereas these birds do not learn to recognize their own eggs, after a few days they do recognize their own chicks and react aggressively towards strange ones. From an adaptive viewpoint this makes sense; eggs cannot wander from the nest site, chicks can. Nevertheless it can rarely be possible for a pair of herring-gulls personally to test the wisdom of this, they probably have an inherited tendency to learn details of the plumage of their own chicks. In this connection it is noteworthy that Tschanz[266] has shown that guillemots, which build no nest but lay their single egg on open cliff ledges, do learn to recognize their own egg. The eggs can roll around and it is clearly adaptive to know which one to retrieve. This may be the reason why guillemot eggs are so variable in ground colour and pattern.

Again, Hinde and Tinbergen[115] describe how young titmice learn to use their feet to hold down large pieces of food in order to break pieces with their bills. Young chaffinches do not learn to do this even when they are reared by titmice foster-parents and this difference is probably inherited. In Chapter 8 we discuss the phenomenon of imprinting which is based upon an inherited tendency in young animals to approach conspicuous objects and learn their characteristics, so that subsequently they are followed and treated as a 'mother figure'.

It seems certain that further investigation will reveal similar inherited predispositions to learn. The extraordinary facility with which bees and wasps learn landmarks around their nests, or polecats learn to seize their prey by the back of the neck are suggestive examples.

Such examples and all that have been described earlier in this chapter give little encouragement for attempts to classify behaviour into that which is inherited and that which is learnt. It may be possible to do this in some cases but it does not seem a very helpful exercise unless it yields information on how the behaviour develops. In the constant struggle to be well adapted, natural selection usually operates on the end result; it matters little how this is achieved. We have already considered some of the factors that may predispose certain groups to rely mainly on inherited behaviour, others on learning. However, even among close relatives there may be wide variation; as mentioned above, the songs of different birds vary widely in their mode of development and this is true even of quite close relatives.

The study of behavioural development is one of fundamental value. The most important property of the nervous system is its ability to store information. Some of this is fed in by genes during the development of the nervous system, other information is added later from the environment and by learning. There seems no good reason to regard these two processes as distinct; they may have almost everything in common. Galambos[80] concludes a paper on mechanisms of learning: 'It could be argued, in brief, that no important gap separates the explanations for how the nervous system comes to be organized during embryological development in the first place; for how it operates to produce the innate responses characteristic of each species in the second place; and for how it becomes reorganized, finally as a result of experiences during life. If this idea should be correct, the solution of any one of these problems would mean that the answer for the others would drop like a ripe plum, so to speak, into our outstretched hands.'

3

External Stimuli

External stimuli can affect behaviour in a number of ways. It is possible to recognize three categories of effect, although they are not sharply separated. Stimuli may *arouse*, they may *elicit* a response and they may serve to *orientate* the animal as it responds. We shall not discuss the orientating role of stimuli any further, but its importance is particularly well exemplified by an animal's use of landmarks as it moves about its environment. Figure 3.1 illustrates one of Tinbergen's experiments on the orientation of the digger wasp, *Philanthus*, to its nest burrow. The wasp learns some characteristics of the ring of pine cones when she leaves the nest and, after they have been moved, the cones orientate—or rather misorientate—her response on a return trip.

Recent physiological work with mammals reveals something of the interaction between the other effects of stimuli—the elicitation of responses and arousal. Every stimulus evokes two types of response within the brain. The first via what may be called 'specific sensory pathways', is one directly related to the stimulus. Visual stimuli evoke activity in the visual centres of the brain, sounds in the auditory centres, and so on. The second type of response is less specific because each incoming sensory pathway also gives off side branches or collaterals which go to a diffuse series of fibre tracts called the 'reticular formation'. This connects via so-called 'non-specific pathways' to all the brain's higher centres and 'arouses' them into action. This means that any stimulus may not only evoke a response pertaining to itself, but also change the animal's state of arousal and responsiveness to other stimuli, both related and unrelated to the first. Parts analogous to the reticular formation of mammals are now being revealed in the brains of lower vertebrates and also in insects.

At the behavioural level we can observe the arousing and motivating effects of stimuli in the phenomenon of 'warm-up' discussed on p. 7, where a response becomes stronger as it progresses. Of course, it may not be just the external stimulus which is arousing the animal, but also the

Fig. 3.1   The digger wasp, *Philanthus triangulum,* builds a nest burrow in sand. Whilst she was in the burrow Tinbergen placed a circle of pine cones around the entrance. When she emerged, the wasp reacted to the new situation by a wavering orientation flight (upper picture) before flying off. Returning with prey (lower picture), she orientated to the circle although it had been moved during her absence. (From Tinbergen,[258] 1951. *The Study of Instinct.* Oxford University Press, London.)

actual performance of the response itself. However, Noirot[201] has shown that the maternal response of mice to a weak stimulus (a dead 1-day-old baby mouse) is stronger if they have previously been given a powerful stimulus (a live baby). She found in some later experiments [201A] that this holds true even when the live baby is presented inside a perforated box so that, although the adult can perceive its presence, it cannot make any appropriate response. The arousal effects of a single 5-minute presentation of the live baby could be detected even when it was several days before the dead baby was presented. This shows that the stimulus alone is sufficient to arouse maternal behaviour and make the mouse persistently more responsive to subsequent stimuli.

Perhaps, following Hinde,[112B] we should say that the maternal responses were 'primed' by the initial stimulus, and retain the term 'arousal' for effects which are more short-lived. Sometimes we know that long-term changes as the result of a stimulus are caused by the induction of hormone secretion. The sight of a male dove courting leads to hormonal changes in the female which make her more ready to take part in nest-building (see Chapter 6).

In other cases motivational changes are linked to stimulus changes through learning. A male bird shows heightened aggressiveness when at home on its territory. Here the male is receiving a variety of stimuli from its surroundings and these have been associated with victory over rivals during previous encounters. As a result the bird is much more responsive to aggressive stimuli than when it strays outside the boundaries of its territory.

With these kinds of interaction between stimuli and motivation in mind, we can discuss some of the problems relating to external stimuli and behaviour.

DIVERSE SENSORY CAPACITIES

Every animal inhabits a world of its own whose character is largely determined by the information it receives from its sense organs. As observers and interpreters of animal behaviour we would be greatly handicapped if we had to rely solely on the evidence of our own senses. Modern instruments enable us to explore the world of animals whose sensory capacities may extend beyond our own. It is here that the study of sensory physiology and of behaviour come to overlap.

For example, the main visual receptors of insects are their compound eyes, whose construction and properties are very different from those of vertebrates. They provide poor image formation, but are excellent for detecting movement and often have a very wide field of view. All insects so far tested prove to have colour vision, but their sensitivity is shifted towards the short-wave end of the spectrum compared with ours. Thus, with a few exceptions, insects cannot see red as a colour, they confuse it with black or dark grey, but on the other hand they can see into the ultra-violet.

We know that foraging bees are initially attracted by the colours of flowers which contrast with the background, but they are aware of contrasts which are invisible to us and vice versa. Bees fly to the flowers of white bryony because the petals reflect large amounts of ultra-violet, although they appear to us as pale green and provide little contrast to the leaves. Further, not all flowers which appear white to us do so to bees. One type of white reflects ultra-violet along with all the other wavelengths and the bees see this as some equivalent of 'white'. The other—equally white to us—reflects all wavelengths except ultra-violet and appears to bees as the colour complementary to ultra-violet, which is blue-green.

Another visual faculty outside our own which bees possess is their sensitivity to the plane of polarization of light. Light which reaches us from areas of blue sky is vibrating predominantly in one plane. The angle of this plane changes in a regular fashion with respect to the position of the sun. Von Frisch[77] has shown how bees can use this to locate the sun's position even when it is obscured directly by clouds or a screen.

Again, the flicker-fusion frequency of the human eye is about 50 per second. This means that a standard filament lamp worked off the A.C. mains supply of 50 cycles per second appears as a steady light source to us. In fact it shows fluctuations of up to 5% of its intensity as the filament heats up and cools down 100 times per second. Insect eyes which may have a flicker-fusion frequency of up to 250 per second can follow these fluctuations, and recordings of nerve impulses coming in from their eyes show corresponding cyclical changes under A.C. lamps. We have no evidence that this fluctuating sensory input affects the behaviour of insects, but it must be borne in mind when performing behavioural experiments under artificial light.

It is not only insects which inhabit a sensory world different from our own; so do much closer relatives. Our world is dominated by sight, but a dog or a cat probably gets more useful information from its sense of smell. The experiments of Griffin[88] and others have shown how bats locate objects and hunt insects on the wing using an auditory echo detection system of extraordinary accuracy. The tropical fish investigated by Lissmann[154] live in water of extreme turbidity where vision is useless. They orientate by setting up an electric field around themselves, using specially modified muscle tissue to generate pulses of electrical energy, and measuring how this field is distorted by other objects in the water.

Information of the type provided by these examples is a necessary preliminary to any thorough behavioural study. However, even if we know an animal's sensory capacities in considerable detail, this alone does not necessarily tell us much about the stimuli to which it will react.

SIGN-STIMULI

In the last chapter we mentioned that it is common to find animals responding only to a part of the stimuli presented to them. They can be taught to discriminate one aspect of a complex situation in this way, but responsiveness to such sign-stimuli is a regular feature of instinctive behaviour in situations where learning can be ruled out. We considered the case of the stimuli which evoke aggressive responses from a male robin on its territory. Here the red breast feathers of the rival are a far stronger stimulus than all the rest of the bird. Pied fly-catcher males have a correspondingly high responsiveness to white feathers—the breast colour in this species. Then there is the classic work of Tinbergen[258] on the sign-stimuli to which a male stickleback responds during its reproductive cycle. Fig. 3.2 shows some of the models which were effective in evoking aggressive responses. They are astonishingly crude. For aggression the red throat of the rival is most important; for courtship responses on the other hand the swollen belly of the female, ripe with eggs, is the most important stimulus.

There are many examples of auditory and chemical sign-stimuli too. Turkey hens which are breeding for the first time will accept as chicks any object which makes the typical cheeping call. On the other hand they ignore visual stimuli in this situation, and deaf turkey hens kill most of their chicks because they never receive the auditory sign-stimulus for parental behaviour.[232A]

Minnows have an extraordinary sensitivity to chemicals from their own species. If a minnow is scratched or wounded in any way so that some of its blood gets into the water, other minnows show panic flight. They show far less fear when the blood of other types of fish is shed. A similar specificity for particular chemicals is shown by many species of moth, whose males are attracted to the females by their scent, which they can detect in minute concentrations.[233]

To label something a 'sign-stimulus' does not imply that a response will never occur in its absence. Some responses, such as that of the male moths, are extremely specific, but a really aggressive male stickleback will attack almost anything, including a ripe female. Even so, we consistently find that the more red there is on a model, the more it is attacked. We are quite justified in calling the red throat a sign-stimulus too.

Similarly, there is often more than one sign-stimulus which will evoke a response. In such cases the lack of one stimulus can be compensated for by an increase to another. Honey-bees guarding their hive entrance detect potential robber bees partly by colour and partly by their characteristic hovering flight. Little balls of wool evoke an attack and brown wool is better than white. However, a white wool lure which is moved in an imitation of a robber's flight works as well as a brown lure held still. Any stickleback model evokes more aggression from a male on his territory if it is held

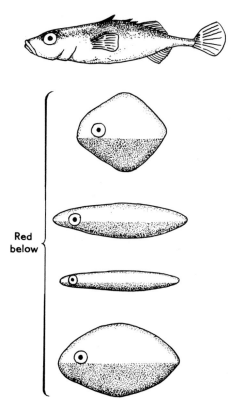

Red
below

Fig 3.2   A series of models used in tests of aggression with male sticklebacks. The four crude imitations coloured red below released more attacks than the accurate model which lacked red. (From Tinbergen,[258] 1951. *The Study of Instinct*. Oxford University Press, London.)

in the head-down threat posture. Here position and red colour can be shown to be additive in their effects. The term 'heterogeneous summation' is used to describe this 'adding up' of diverse stimuli. It may be considered as an extension of the stimulus summation already described for reflexes and complex behaviour in Chapter 1.

Selective responsiveness to sign-stimuli is clearly of great adaptive significance in the lives of many animals, particularly those which rely primarily upon inherited behaviour. Sign-stimuli will usually be involved where it is important never to miss making a response to the stimulus, but a few false responses do not matter much. Driving away rival males from its territory is so important to a male stickleback that it must show extreme responsiveness to red. Nearly all the red objects which it sees will be rival males, but occasionally red petals from flowers may fall into the water and cause a male to waste some time in futile aggression. Tinbergen describes how a male he was observing made a threatening display towards a red post-office van which drove past the windows of the aquarium. Such rare false positives do not count for much when set against the advantages of being consistently aggressive towards other males.

Again animals must never fail to respond to the sign-stimuli provided by a predator, or by the alarm calls of other individuals. It is advantageous to respond to the alarm calls of other species too, so that no matter which first perceives danger, all are warned. Marler[179] discusses the characteristics of an 'ideal' alarm call which must carry as far as possible whilst giving the least chance of a predator being able to locate the calling bird. Among other things, this means the call should be of constant pitch and have a gradual beginning and ending. The alarm calls of many finches, buntings and thrushes are remarkably similar and it seems probable that they have evolved towards each other so as to provide a common sign-stimulus to which all respond (see further discussion in Chapter 7, p. 171 and Fig. 7.7).

Another way in which many small birds are protected against predators is by possessing a strong flight response from any animal having large staring eyes, such as are characteristic of most birds of prey and owls. Blest[30] has shown that small birds are far more frightened by a simple pattern resembling an eye which is flashed at them, than other patterns with equal amounts of contrast and outline. Many different types of butterfly and moth, which are themselves potential prey for small birds, have developed eye patterns on their wings. Sometimes these are of amazing detail (see Fig. 3.3) and whilst ordinarily hidden they are suddenly flashed into view when the moth is touched. There can be no doubt that moths have been able to develop this mechanism because of the strong selection for small birds to respond to the eye-pattern sign-stimulus. It is better to be frightened away by a harmless moth occasionally, than ever to fail in the response to a predator.

In some contrast to dependence upon sign-stimuli, recognition can be based on a complex of features. We can illustrate the latter by considering the way in which we recognize a particular human face. Usually no one feature is a sign-stimulus, but almost every aspect combines to produce a unique pattern. It takes time to learn this and it is clearly influenced by our

culture. We find it more difficult to distinguish Mongoloid faces than
Caucasian ones; and the Mongoloid races apparently have a similar
difficulty with Caucasian faces.

There are some situations among animals which emphasize this difference between sign-stimulus and 'complex' recognition. The males of

(a)

(b)

Fig. 3.3 The moth *Automeris coresus* (a) at rest, and (b) displaying the vivid
eye-spots on its hind wings in response to a light touch. (Modified from Blest,[31]
1957, *Behaviour*, **11**, 257.)

various species of ducks have strikingly different and boldly patterned plumage. Conversely the females resemble one another much more closely because they are cryptically coloured in mottled shades of brown. Evidence is now accumulating that females have an inherited ability to recognize their own type of male, but males have to learn to recognize the subtle pattern of their females. They complete this learning within a few weeks of hatching, using their mother as a model (see p. 162 for more details). Seitz, (see Baerends & Baerends-van Roon[11]) studied the tropical fish *Astatotilapia* which has brightly marked males and dull females. He found a situation somewhat analogous to that of the ducks. Females have an inherited response to sign-stimuli provided by the males, but males learn details of the female markings. This means that whilst females will readily respond to male models, only inexperienced males will court simple models of females. Once they have sexual experience they respond only to the whole complex of female markings.

In our discussion of sign-stimuli we have constantly used examples of responsiveness to the particular sounds, scents or colours of another animal. It was Lorenz who first developed the idea that such characters are in fact specially evolved to evoke such responses. He called them *releasers*, and pointed out that the releaser and the response from the animal which 'receives' have become mutually adapted to each other. They form a signalling system which has often become elaborated into a primitive language in which the effect of a releaser is enhanced by a display movement. Cockatoos erect their brightly coloured crest feathers whilst honeybees open the Nasanoff gland on the abdomen in order that its scent may be dispersed. We shall refer to releasers in other parts of this book because their evolution is linked closely to the evolution of courtship and threat displays. Until their signal function was understood, many of the bright colours of such animals as birds and butterflies were hard to account for. Now it is often possible to deduce quite accurately the type of display a bird or a fish will use simply from seeing its colour pattern and shape. Fish with colour markings on the gill covers will tend to have elaborate frontal displays in which the covers are spread, and so on (see Fig. 5.3a).

The central chapters of Marler and Hamilton's book[179B] provide an impressive survey, covering the entire animal kingdom, of the diverse ways in which sense organs, behaviour and releasers have become integrated into a communication system adapted to the requirements of each species. Most releasers have evolved for intraspecific communication, but not all. The eye-spots on the wings of moths have developed because they frighten birds. The colours and scents of flowers have evolved because they attract bees and other pollinating insects.

The recent work of ethologists has provided some beautiful analyses of the sign-stimuli which evoke particular responses. One of the most interesting findings is that different aspects of the same object may be effective according to the type of behaviour which is involved. For example the answer to the question, 'How does a gull recognize an egg?', depends on whether the bird in question is incubating eggs or feeding on them. If incubating, then the experiments of Baerends[10] with herring-gulls show that the most important sign-stimulus is speckling. The more spots an egg has and the more they contrast with the background, the more strongly is the gull stimulated to roll it into the nest and incubate. The egg's background colour is far less important, although green is best. Shape, on the other hand, is ignored; square or cone-shaped eggs are accepted perfectly well provided they are speckled. Contrast this with the situation when a gull is searching for eggs as food. Now shape and 'wholeness' are the most important features by which it recognizes them.

Tinbergen and his co-workers[264] have been able to analyse how black-headed gulls recognize egg-shells and distinguish them from eggs. They remove the egg-shells from the nest as soon as the chicks have hatched because empty shells attract predators. The gulls recognize a shell mainly because it shows a thin white edge which is serrated and 'hollowness' by itself is not effective. Fig. 3.4 illustrates some of the models used in their tests.

(a)   (b)   (c)   (d)

Fig. 3.4 Some of the egg-shells used to determine how a gull recognizes an empty shell. The real shell (a), 'painted rim' (b) and 'notched rim' (c) elicited more carrying than 'smooth rim' (d) which was rolled back into the nest. (Modified from Tinbergen et al.,[264] 1962, *British Birds*, **55**, 120.)

'Supernormal' stimuli

Ethologists usually investigate sign-stimulus situations by making models of the stimulus object and changing parts of it in rotation to see which are most important. Not uncommonly they have found that it is possible to produce a model which is 'supernormal', i.e. it evokes a stronger response than does the natural object. The most striking examples have again come from the incubation behaviour of birds. Tests have been made with the herring-gull, the grey-lag goose and the oyster-catcher and in all three the larger an egg is, within broad limits, the more it stimulates incubation. This results in the bizarre situation illustrated in Fig. 3.5 where an oyster-catcher

Fig. 3.5 An oyster-catcher attempting to brood a giant egg in preference to its own egg (foreground) or a herring-gull's egg (left foreground). The bird's original nest site was equidistant between the three 'test' eggs. See also Fig. 2.6. (From Tinbergen,[258] 1951, *The Study of Instinct*. Oxford University Press, London.)

invariably chooses a giant egg in preference to its own as an object for brooding. The same species chooses an artificially large clutch of five or six eggs in preference to its own clutch of three.

Herring-gull chicks will peck at models of the parent's bill from which they are normally fed with regurgitated fish. The herring-gull's bill is clear yellow with a red circular patch on the lower mandible (see Fig. 3.6). The artificial bill, also shown in Fig. 3.6, is thinner than the real one, coloured

Fig. 3.6 (a) Accurate, 3-dimensional model of a herring-gull's head, and (b) 'supernormal' bill. The latter received 26% more pecks from young chicks. (From Tinbergen and Perdeck, [265] 1950, *Behaviour*, **3**, 1.)

red with three white bars at the tip; this attracts more pecks from naïve chicks than does a realistic copy of the natural bill and head.[265]

A further supernormal stimulus was revealed by the work of Magnus[171] with the silver-washed fritillary butterfly. Males are attracted to females by the flashing orange wing pattern as they fly by. Magnus could attract males to a revolving drum which 'flashed' a wing pattern at any required speed. The normal wing-beat frequency of a female fritillary is about 8 per second, but males show stronger responses the faster the wings of models are made to beat, up to as high as 75 per second.

In all these examples there is no difficulty in understanding why the natural stimulus has not evolved further towards the supernormal condition. Female fritillaries would attract more males if they beat their wings faster, but wings are not only for this purpose, they have to move insects through the air and there are severe mechanical limitations on their speed. Similarly a herring-gull's bill would probably be highly inefficient in all but attracting the pecks of its chicks, if it were as long and narrow as the supernormal model.

There remains the converse question of why the response of the receiver is not perfectly adjusted to the natural situation. We do not know the answer to this. Perhaps the mechanism which provides the selective responsiveness to a sign-stimulus has no upper limit. An increase in the 'quantity' of sign-stimulus inevitably results in an abnormally strong response. Natural selection can rarely have an opportunity to operate against this kind of system. On the other hand the important thing about supernormal stimuli may not be an increase to the 'sign' aspects but the fact that they are more conspicuous or arouse the animal more in a general way, so that its response is stronger. This seems likely to be the explanation for supernormal female models in the fritillary. Bees and other insects are known to approach flickering lights and within limits the faster the flicker, the more attractive. For the same reason bees prefer objects such as stars with broken outlines to circles or other plain shapes. As a bee flies over a broken shape the images of its margins will pass across the separate ommatidia of the bee's eye producing the effect of a flickering light source. The fast-moving flicker of the fritillary model will work in the same way.

STIMULUS FILTERING

In all our discussions of sign-stimuli it has been implicit that the animal selects one part of the environment to respond to and ignores the rest. This immediately raises the question of how far the ignored stimuli penetrate. Where is the 'filter' which separates them from the sign-stimuli and renders them ineffective?

Before we discuss this problem it is worth pointing out that no animal's brain can ever handle more than a fraction of the potential information that comes from its sense organs. Sense organs transform some physical feature of the environment—a colour, a sound, a smell—into a series of nerve impulses which travel along the sensory fibres up to the brain. Barlow[12] calculates that there are roughly 3 million such fibres entering the human brain. If each of them is regarded as a switch which can be either 'off' or 'on' then $2^{3 \text{ million}}$ different combinations of input are possible. To deal with them the brain has 10^{10} neurons and, vast as this latter number is, it is paltry in relation to the former which is far larger than Eddington's figure (10^{79}) for the number of particles in the entire universe!

Obviously not all of the $2^{3 \text{ million}}$ combinations of sensory fibre activity can occur, and anyway a large number of the combinations may be meaningless as far as the brain is concerned and can be ignored. Barlow suggests ways in which the brain can further economize on handling sensory information: some form of filtration is universal.

As applied to sign-stimulus situations filters can operate either peripherally or centrally, and we can best explain these terms by examples.

Peripheral filtering

Quite obviously what an animal responds to is limited in the first place by its sense organs. We have already discussed the capacities of sense organs and often they act as filters. Our own eyes filter off all ultra-violet light and our ears transmit no sound with a frequency higher than 20 kilocycles. Sometimes selective transmission by sense organs produces the effect of a sign-stimulus. When feeding, frogs respond much more strongly to small, dark, circular objects moving close to them than to movements of large objects or the whole background. This makes sense in an animal whose chief diet is flies. Lettvin et al.[150] have shown that the light receptors of the frog's retina are connected together so as to form 'receptor fields'. Some of these fields are specialized to respond preferentially to the intermittent movement of small, dark, convex objects and Lettvin and his co-workers appropriately call these receptor fields 'bug detectors'. The responsiveness of the frog to such stimuli is produced by peripheral rather than by central filtering.

There are several good examples of a similar type from insects. Roth[220] describes how male mosquitos respond with high selectivity to the sound of their females' wings which beat at a characteristic frequency, different from their own. It is the delicate hairs on the male's antennae which respond to the sound. Roth suggests that their physical structure is such that they vibrate most readily at the female's wing-beat frequency, just as a tuning fork will vibrate if its own frequency is sounded near by.

The sense organs also determine the amazing selective responsiveness which some male moths have for the scent produced by females. Males of the silk moth will gather from considerable distances when a virgin female is put out in a cage; once she has mated she ceases to produce the scent. It has been possible to isolate the female scent substance and define it chemically. A few synthetic substances which are closely related to female substance will attract males, but no others. Subsequently Schneider[233] has found that the scent receptors on the male's antenna are extremely sensitive to female substance, but show little or no response to any other chemicals.

Central filtering—the innate releasing mechanism

In contrast to these examples there are many where an animal's responsiveness to a sign-stimulus cannot easily be ascribed to its sense organs. As we have seen, sometimes animals respond to different aspects of the same situation depending on their 'mood'. Here we must assume that some part of the selection is taking place centrally; i.e. the brain is picking out certain fractions of the stream of information from the sense organs and using them to direct a certain type of response.

Lorenz[155] suggested that a special mechanism was responsible for stimulus filtering and proposed the term 'Innate Releasing Mechanism' (IRM) to describe it. An IRM is defined by Tinbergen[258] as a '. . . special neurosensory mechanism that releases the reaction and is responsible for its (the reaction's) selective susceptibility to a very special combination of sign stimuli'.

It will be seen how closely the IRM fits in with the concept of the 'releaser'. IRMs and releasers are mutually adapted to one another. One animal gives a stereotyped signal by its releaser to which another animal is especially responsive via its equivalent IRM.

As defined by Tinbergen the IRM concept does not exclude the possibility of peripheral filtering, but there can be no doubt that ethologists used to think primarily in terms of central filtering. Perhaps some of the trouble lies in the 'mechanism' part of the term IRM. This might be taken to imply that there is one site in the nervous system where the filter operates and that this is similar in all cases. One might equally object to the word 'releasing', because this implies that the stimulus merely triggers off a response which is ready-primed and ignores the arousing effects of all stimuli.

However we may with reservations retain the term if it proves useful and provokes research into the exact nature of stimulus filters. In fact although we have several excellent analyses of the stimulus situation to which a particular IRM is adapted, we know next to nothing about how such a filter may work. Before this problem can be approached we need to define very exactly what are the characteristics of particular IRMs. Thus among other things, we must determine just how 'central' the filter is or whether some of the sensitivity to key stimuli is due to sense organs after all. We will examine two examples where analysis of an IRM has gone some way.

The pecking response of gull chicks*

The young chicks of gulls and terns solicit feeding by pecking at their parents' bills. Tinbergen and Perdeck's[265] work on the stimuli which evoke the herring-gull chicks' pecking has already been mentioned. The adult's bill is yellow with a red patch on the lower mandible (see Fig. 3.6). Tinbergen and Perdeck made a number of cardboard models in which they varied (i) head colour, (ii) head shape, (iii) bill colour, (iv) colour of patch on lower mandible and (v) degree of contrast of patch (black through shades of grey to white) on a medium-grey bill. Fig. 3.7 shows the results of series (iii), (iv) and (v) diagrammatically. In all these tests they took care that the models were presented to the chicks in standard manner. Each type of model was presented first to equal numbers of completely naïve chicks, i.e. those which had never seen a model or a real gull's head before; thereafter the order of presentation was randomized.

In summary, Tinbergen and Perdeck found that the mandible patch was the most important aspect of the parent's head. It worked partly by contrast (grey bills with black or white patches got more responses than plain bills), and partly by virtue of its red colour. A plain red bill is more attractive than any other colour, and a red patch more effective than a black one even though its contrast may be lower. The colour or shape of the head beyond the bill are immaterial. There are, in addition, other features of bill shape and movement which we will ignore here.

We can now begin to specify some of the characters of the IRM which filters out the sign-stimulus in the normal situation. It selects redness and contrast but eliminates bill colour and stimuli from the parent's head.

Weidmann and Weidmann[276-7] have repeated these experiments using the black-headed gull and Hailman[92] has used the laughing-gull. Both these species have a plain red bill and in both cases, as with the herring-gull, the naïve chicks' preference for red was strong. The same is true of kittiwakes and arctic terns which also have red on the bill or inside the mouth.

There is plenty of good evidence that birds have full colour vision similar to our own, therefore it is tempting to consider the red on the parent's bill as a releaser and the chick's corresponding responsiveness as being one property of the pecking IRM, the two being mutually adapted.

However this preference for red might not be due to a central filter. In the retinal light receptors of gulls and many other birds, there are abundant droplets of red, orange and yellow oil. Incoming light has to pass through these droplets before it reaches the light-sensitive pigment. Light at the

* There is a very full discussion of this problem in Hailman[93A].

Fig. 3.7 Three series of model heads used in the pecking tests: (a) measures the effect of bill colour, (b) that of patch colour (all the bills were yellow), and (c) the effect of varying the contrast between patch and bill colour (all the bills were grey). The length of the bar beside each model is proportional to the number of pecks it received. (From Tinbergen and Perdeck,[265] 1950, *Behaviour*, **3**, 1.)

blue end of the spectrum will tend to be filtered out whilst red light will pass through. This does not mean that birds cannot see blue, but merely that it will appear darker than an equivalent intensity of red light. Perhaps then part of the properties of the IRM are in the gull chick's retina.

Comparative evidence shows that whilst this may be a sufficient explanation in those species which prefer red, it is not the whole story. Two species of tern have been studied in which the young chicks respond more strongly to black bills than to red (sandwich tern[277]) or find black equally attractive (wideawake tern[58]). It cannot be simply coincidence that the adults of these two species have black bills; that of the wideawake is plain black, whilst the sandwich tern's bill has a yellow tip. There seems to be no possible property of the retina which could make black appear as bright red provided, as in these tests, precautions are taken to equalize contrast with background. The IRM in these two terns probably includes a truly central component which responds selectively to black.

The universality of oil droplets in the retina of birds makes it likely that the red bills of the great majority of gulls and terns are adapted to elicit the maximum strength of response from their chicks who see red as the brightest colour. For the same reason flowers which are pollinated by birds are nearly all a vivid red colour. A number of other factors in the feeding behaviour of the adult, such as the angle at which it holds the bill and the way it is moved back and forth, also serve to focus the chick's attention on the bill tip, where food will be presented.

With the majority of gull species we do not need to ascribe any filtering properties to the IRM beyond those of the retina itself. The chick pecks towards the most conspicuous object in front of its eyes. The irrelevance of the parent's head, both in colour and in shape, is probably because it is right at the edge of the visual field of a chick whose attention is focused on the bill tip. Weidmann[277] considers that black-headed gull chicks cannot see the parent's head at all clearly when they are pecking.

With the black-billed terns, the mutual adaption of releaser and IRM have proceeded one stage further. For some reason selection has not favoured the retention of red bills and the chick's responsiveness has followed the change to black.* Here there may be a central component selecting black in the chicks' IRM, but presumably because of the basic retinal characteristics red still remains very attractive to them and second best to black.

There is some other evidence that central filters which are not affected by experience may be involved in the colour preferences of young birds.

---

* At least this is the simplest explanation. When deducing evolutionary trends without fossil evidence one must always allow for a trend proceeding in the opposite direction. Ancestral gulls and terns may **all** have had black bills and only these two species have retained them. In which case the argument, somewhat tortuously, must run backwards too.

Kear[133-4] placed newly-hatched chicks of a number of precocial birds (those which can run about and peck after hatching) in an arena where they could peck at coloured spots. Most species showed strong preferences though they were never rewarded for pecking at any particular colour. She confirmed the attractiveness of red for gull chicks. Moor-hen and coot chicks also preferred red and both these species are fed from the parent's bill at first. The correspondence is good in the moor-hen which has a red bill, but although the coot's bill is vivid white, young coots still prefer red. Pheasant chicks and ducklings of several species all pecked most strongly at green, although green was very unattractive to the other birds tested. It is possible that this preference for green may simply be a response to brightness because the eyes of ducks and pheasants are more sensitive to green wavelengths. Kear[134] did find that varying the brightness of the colour spots affected the relative popularity of some colours. However it seemed most likely that ducklings do select green by its hue and it may be relevant that pheasant chicks and ducklings have to feed themselves as soon as they leave the nest and mostly eat green plants of various kinds.

Oppenheim[202A] has recently examined the colour preference of mallard ducklings in great detail. Like Kear he found a strong preference for pecking at green, and this seems to be because the ducklings are attracted by green objects which contrast with their background and not just because they move towards any area of green colour. Oppenheim found that the initial preference for green is unaffected if eggs are incubated and the ducklings hatch in total darkness, or if he opened a window in the shell and illuminated the eyes of the embryo with white or yellow light for many hours before hatching. He was unable to account for the green preference by the properties of the duck's retina alone.

All these observations suggest that in some cases colour preferences have an inherited basis which is 'built in' to the central nervous system. The next example examines selective responsiveness to other aspects of visual perception—shape and movement.

The alarm response to a flying predator

Many ground-nesting birds such as ducks, geese, pheasants and turkeys give alarm calls and crouch when a bird of prey passes overhead. This response is particularly strong in a female bird with chicks. Turkey hens spread their tails as they call and the chicks run to shelter beneath them.

Tinbergen[258] describes some experiments which Lorenz and he performed in 1937. They constructed a number of simple cardboard silhouettes of flying birds which they caused to 'fly' on a wire across a pen containing geese or various game-birds.

They recorded the presence and strength of alarm responses and found that the most effective models had the common feature of a 'short neck' which is characteristic of most birds of prey. The most striking experiment which has become something of a classic, albeit a controversial one, used the silhouette shown in Fig. 3.8. This evoked no alarm if flown to the left with the long neck leading (simulating a goose), but did evoke alarm when the short neck led, (simulating a hawk). Here then is a sign-stimulus situation in which it is not merely shape which is effective, but the relationship of shape to movement. It is unlikely that such selective responsiveness can be due to properties of the bird's retina alone.

Fig. 3.8 The 'hawk-goose' silhouette used in experiments on the alarm response of geese and game-birds. (From Tinbergen,[258] 1951, *The Study of Instinct*. Oxford University Press, London.)

A number of critics have been sceptical of the suggestion that a response to such a complex stimulus configuration can be inherited. Schneirla[233A] has suggested that there is a far more general explanation of such responses. All animals from Protozoa upwards, he argues, tend to approach any sources of moderate stimulation—sound, light or chemical—and to withdraw from strong and suddenly increasing stimulation. He suggests that the silhouette flown in the 'hawk' direction is more alarming simply because it has a broad leading edge. This will cause a more abrupt darkening of the bird's visual field than the 'goose' silhouette which broadens gradually from a tapered leading edge. Schneirla discounts any IRM specially adapted to the hawk silhouette, and predicts that one would get the same differential effect from a plain triangle depending on which way it was flown. Base leading, he suggests, will induce more alarm than apex leading.

Certainly we must examine critically any claims to have demonstrated inherited specific responsiveness to complex stimuli and the original experiments are open to a number of criticisms. In the first place nothing can be argued about the nature of an inherited response to a 'hawk shape' unless naïve birds are used. Lorenz and Tinbergen's birds could have learnt previously what hawks look like. Secondly, they scored only the reaction of the group, and individuals may vary and be influenced by the behaviour of others.

The first attempt to apply more strict criteria was that of Hirsch, Lindley and Tolman[117] who used newly-hatched white leghorn chicks and scored their responses individually. They found no greater alarm to the hawk model than the goose. Both proved moderately alarming when first flown overhead, but the chicks rapidly got used to them and took little notice.

Tinbergen[262] and others have quite reasonably suggested that it is difficult to generalize from white leghorns to wild geese. The former may have lost any inherited response to hawk shape from many generations of domesticity with no selection in its favour. More recently Melzack et al.[184] have repeated experiments of this type with wild mallard ducklings. They could confirm Lorenz and Tinbergen's original result to the extent that roughly twice as many naïve ducklings showed fear responses to the hawk when first flown over, as to the model reversed and flown as a goose. However, like Hirsch et al., they found that the ducklings rapidly ceased to show fear to either model, although they never relaxed their vigilance and always followed with their eyes any model that was flown. Green et al.[86A,86B] got the same result—the hawk elicited much more alarm than the goose from naïve ducklings which had no prior experience of overhead objects. In addition, they tested Schneirla's prediction concerning plain, triangular silhouettes and could not uphold it. The ducklings showed very low levels of alarm to the triangle whichever way it was flown.

However a further complication is added by the work of Schleidt[232] with turkeys. He finds that he can release the alarm response with models of almost any shape—a plain circle for example—provided only that it flies over at the correct speed. This is rather slow, between 5 and 10 model-lengths per second and corresponds well to the apparent speed of birds of prey in flight. A tiny model only an inch long will evoke a response if it moves at this speed. Schleidt found that turkeys would rapidly cease to respond to any model with which they were familiar, but novel shapes, no matter how they moved usually frightened them. He suggests that Lorenz and Tinbergen's geese did not respond to long-necked models because they were already accustomed to the sight of geese flying over, but this cannot explain Melzack's or Green's results.

It is rather difficult to equate the results of all these experiments with the natural situation. Birds cease to respond to familiar models no matter how hawk-like, but in the wild they go on responding to birds of prey although few will ever have been attacked. Perhaps this is because the models always appear in the same place and in the same way. Also, as Schleidt points out, wild birds are rarely confronted with flying hawks at the high frequency with which experimenters present hawk models! The scarcity of hawks must be a powerful factor in maintaining responsiveness.

In summary, there is evidence that wild birds do possess an IRM which enables them to respond to birds of prey on the first occasion that they see them. This IRM probably has different properties in different species but short neck and relative speed of movement are among them. Peripheral filtering is unlikely to be involved here.

Any attempt to analyse the properties of an IRM is bound to involve work on both sensory physiology and behaviour. Clearly the IRM, if it has any use as a concept, must be taken to include the whole range of possible filters beginning at the sense organs themselves. The end result of the filtration process is to adapt the animal to respond selectively to some sign-stimulus from the environment, which may be a specially evolved releaser whose properties are mutually adapted to those of the IRM. The main unsolved problem concerns the nature of central filters. It seems unlikely that these rely on any single type of mechanism, although the properties of central filters which mediate learnt discriminations and those such as we have dealt with here, whose development is controlled by genes, probably have much in common.

The experiments of psychologists investigating how an animal recognizes objects, show that it does not respond to all their features simultaneously. Rather it directs what may be called a central 'attention mechanism', first upon one set of cues—say 'brightness', 'colour' or 'shape'—and then upon another.[82A] Clearly such an attention mechanism shares many properties with the central filters we have been discussing. Ethologists have been impressed by the manner in which a change to an animal's 'mood' causes a switch in attention; one filter is inhibited and another is brought into operation. When he first emerges from the pupa a male grayling butterfly responds most strongly to blue and yellow which lead him to flowers. Later, using the same sense organs he responds preferentially to brown when courting a female. In some way the filter changes as his 'mood' changes from feeding to courting.[258]

With this example we conclude this chapter as we began. It demonstrates how the stimuli which are effective depend on how an animal is motivated. We must now turn to consider the nature of motivation.

4

Motivation

It is a common observation that the same stimulus given to the same animal at different times does not always evoke the same response. Something inside the animal must have changed and we invoke an 'intervening variable'. This is something which comes between two things we can measure—in this case the stimulus we give and the response we get out— and affects the relationship between them. We must admit that in some cases we know next to nothing about the real nature of such variables, and some groups of behaviour workers refuse to use them and concentrate entirely on directly observable aspects of behaviour. However, most people who have worked with animals under fairly natural conditions recognize the necessity to invoke intervening variables in behaviour. At this stage of our knowledge it is profitable to define them carefully and investigate their properties. Already in this book we have mentioned two factors with different characteristics which alter the relationship between a stimulus and the response it evokes. These were 'fatigue' (p. 8) and 'maturation' (p. 24). To these we may add two others: 'learning', which will be dealt with in Chapters 8 and 9, and 'motivation' which concerns us here.

Changes in 'motivation' are deduced when we can eliminate the other factors just listed, but still observe that an animal spontaneously changes its behaviour or shows a changed threshold to particular types of stimuli. Sometimes a very slight stimulus may be adequate to evoke a powerful response, at other times far stronger stimuli are ineffective. As mentioned in Chapter 1, this is one characteristic which separates reflexes from more complex behaviour, because the former tend to have a consistently low threshold. With complex behaviour it is common to find that it is not just one stimulus which is effective or ineffective, but a whole range of stimuli which are related functionally to one another. Thus an animal's thresholds of response to all stimuli connected with food and feeding behaviour will rise and fall together; so will those connected with sexual stimuli, and so on. Because of this, many workers and particularly ethologists have regarded motivational changes as highly specific, tending to make the animal respond in a particular way. The result of such a tendency is that an animal's behaviour becomes organized so that it achieves a particular goal. A dog kept for many hours without food moves about restlessly; it is highly responsive to the smell or sight of food or any other stimulus, such as the sight of a food bowl or the sound of a knife being sharpened, which it has learnt to associate with food. It is not to be diverted by a bowl of water or a ball to play with; its restlessness continues until it has eaten.

However it may be misleading to think entirely in terms of specific motivational states. As we mentioned in the last chapter, any stimulus may produce a non-specific arousal via the reticular formation of the brain. This will render the animal more responsive to a wide range of stimuli and we might describe this change as a rise in 'general motivation' or, as it is called by many psychologists, 'general drive'. Although it might seem a straightforward question, it is in fact very difficult to collect really conclusive evidence as to whether motivation is general or specific; there is a good discussion of the problem in Chapter 9 of Hinde,[112B] (see also Grossman[88C]). In what follows, we shall assume that there is a considerable degree of specificity, although it is certainly wrong to think of motivation as being rigidly compartmentalized. Different motivational systems interact with one another, and in the next chapter we must consider how such interactions affect behaviour.

A specific motivation is often called a 'drive', thus the food-deprived dog in the example given above might be described as having a high feeding drive. This term needs to be used with care because we cannot measure drive directly. Usually we measure only the animal's response to various types of stimuli. If we find that the dog is highly responsive to food stimuli it might be preferable to say that it shows a high feeding tendency, or a low feeding threshold because this is what we actually observe. 'Tendency' is a useful term in motivational studies and will be employed here wherever possible.

In ourselves, specific motivational states are often associated with strong subjective feelings or emotions but we cannot tell whether animals feel emotions in the same way. All we can say with confidence is that there are various physiological changes in the body correlated with emotion and that animals show these. The mouth becomes dry, sweating starts, heartbeat accelerates and so on. These changes mostly follow the secretion of adrenalin into the bloodstream and are discussed more fully in Chapter 5. Although the emotions of anger, fear and lust feel subjectively very different, they share many features of this physiological arousal which prepares the body for violent action of any type.

Since we cannot measure emotion in animals, we usually have to deduce how an animal is motivated by observing how it behaves. Here we inevitably introduce a subjective element because we have to identify motivation with reference to a function that seems reasonable to us. If one animal attacks another we ascribe this behaviour to aggressive motivation; if it eats we ascribe this to feeding motivation and so on. Clear-cut cases such as this offer little difficulty if treated with common sense, but it must be admitted that not all behaviour is so easily interpreted.

SOME CHARACTERISTICS OF GOAL-ORIENTATED BEHAVIOUR

We have already mentioned that specific motivation leads to an animal orientating its behaviour towards a specific goal. There is no need to postulate that animals consciously strive to achieve their goal; they may do but we shall probably never be able to prove it one way or the other, and the question is not a helpful one for the experimental analysis of behaviour. Quite often we can observe three stages in the animal's behaviour.

1. A phase of searching for the goal.
2. Behaviour orientated around the goal once it is found.
3. A phase of quiescence following the achievement of the goal.

1. The searching phase is usually called the phase of '*appetitive behaviour*'. This is best described with relation to feeding from which it probably took its name. A hungry animal can rarely just get up and eat; it has to seek out food and the behaviour patterns it employs may be many and diverse. This is particularly conspicuous with predatory animals which actively hunt their prey. Their hunting behaviour is variable and greatly modified by their previous experience.

It is often impossible to identify the nature of appetitive behaviour until we observe the actual goal. Animals searching for food, water or a mate may behave very similarly, but in each case the stimuli required to bring the search to an end are highly specific.

2. When the appropriate goal stimuli are located, the animal's behaviour changes. The variable searching patterns now give way to a series of responses directed at the goal which are often stereotyped, fixed action patterns. They are called *consummatory acts*; eating is the consummatory act of feeding behaviour, drinking relates to thirst, copulation to sexual behaviour, and so on.

3. Consummatory acts are normally followed by a period of quiescence when the animal is no longer responsive to stimuli from the goal and

shows no further appetitive behaviour. This quiescence relates, of course, only to one type of behaviour; the animal may be actively pursuing some other goal. In some cases responsiveness slowly builds up again so that there is a fairly direct relationship between threshold to goal stimuli and time since the last performance of the consummatory act.

Whilst appetitive behaviour and consummatory act, particularly the former, are still useful as descriptive terms, there is no question of their being rigid categories which can be applied universally.

For example, it is not easy to classify the nest-building behaviour of, say, a blackbird in this way. The blackbird may begin by searching for large twigs to form a foundation. When this is complete the sides are made from finer material; then mud is collected and shaped to form the cup and finally this is lined with fine grass and hair. We might classify this behaviour by postulating that, whilst the completed nest is the ultimate goal stimulus, there are a series of 'sub-goals'—nest foundation, sides, cup, etc.—along the way, each with its own appetitive behaviour and consummatory act.

Even in feeding behaviour, which in some cases fits the scheme very well, the pattern varies. A horse grazing in a field literally has its food at its feet. The appetitive behaviour phase is short or non-existent and the consummatory act of feeding may continue without a break for an hour or so until the animal is full. This situation is totally different from the feeding behaviour of a titmouse picking minute insects off leaves. Here each brief consummatory act is not followed by quiescence, but by a further phase of appetitive searching. Eventually after some hundreds of such sequences, appetitive behaviour ceases.

The existence of an identifiable goal and appetitive behaviour directed towards locating it has sometimes been used as evidence for postulating specific motivational states or drives. Some drives are labelled 'biogenic', or related to an urgent biological need. Everybody would agree that feeding and drinking come into this category and, as we have seen, they best fit the appetitive behaviour/consummatory act/quiescence scheme outlined above. Many would call sex a biogenic drive too, and it also fits the scheme fairly well, at least in vertebrates, although it is remarkable how much sexual arousal can be affected by the external stimulus situation in some cases— far more than is possible with feeding or drinking. For example Hale[93B] has shown that even apparently trivial changes to the female test animal were sufficient to arouse a sexually exhausted bull to mount once more. Ethologists commonly postulate attack and escape drives, and in the next chapter we shall discuss these in relation to territorial behaviour. If we exclude predatory animals which attack their prey as part of their feeding behaviour, animals do not normally move around actively searching for things to attack or to escape from and both types of behaviour are far more dependent on external stimuli than are feeding and drinking. The tendency to attack does fluctuate—testosterone is one of the factors that affects this—but in addition it is rapidly aroused by certain types of stimuli and does not show a simple relationship between threshold and length of time since the last attack. Escape certainly does not fit into the standard scheme; it is almost exclusively under the control of external stimuli. Apart from this control, there is no evidence that the tendency to escape fluctuates save as a result of learning. Animals quickly cease to respond to stimuli which, though alarming at first, are not associated with punishment (e.g. scarecrows) but novel stimuli may once more evoke strong escape. It would be highly inadaptive if this were not the case.

At one time or another, sleep, parental behaviour and exploratory behaviour have all been ascribed to specific drives. Sleep, which we now know to be associated with a particular pattern of neural activity, not inactivity, superficially fits the appetitive behaviour/consummatory act/quiescence scheme rather well. However, little critical analysis in these terms has been made for sleep or any of the other suggested drives.

Experimental psychologists often use the term 'drive' in a rather different sense when they refer to 'secondary' or 'learnt drives'. (A 'primary drive' in their terminology is equivalent to those we have just been discussing.) Miller[187] says, 'Thus, if a child that has not previously feared dogs learns to fear them after having been bitten, it shows that fear is learnable.' From the biological point of view it is more appropriate to concentrate on the fact that it is the *stimulus* which is learnt, and becomes associated with the escape system.

WHY POSTULATE DRIVE?

So far we have been using the drive concept in a descriptive way without comment. We have seen that what might be called the 'classical picture' of a drive in operation is rarer than might be supposed from the frequency with which the term crops up in ethological writings. In particular, it is difficult to find good examples from the behaviour of invertebrates. We must now consider in more detail exactly when and why we need such a concept.

The trouble with a term like 'drive' is that it can take on a substance which we cannot really justify. It may be stretched to account for a number of different behavioural phenomena which are thus given a kind of false unity. Hinde[110] discusses this problem and lists no fewer than six phenomena which drive has been used to explain. We can consider three of these in more detail.

Fluctuations in responsiveness

We have already dealt with the general nature of this phenomenon. In some cases it is easy to relate the fluctuating responsiveness to regular changes in an animal's physiological state. One only needs to know how long it is since an animal fed or drank to get a good idea of how responsive it will be to food or water respectively. Feeding and drinking provide the clearest examples of behaviour forming part of a homeostatic system. Some part of the system detects shortage of water and appetitive behaviour and responsiveness to water increase until it has drunk. Thereafter responsiveness falls until water begins to run short once more.

Sexual responsiveness also shows great fluctuations, especially in animals with a well-marked breeding season which show no sexual behaviour at all for much of the year. It is often possible to demonstrate a good correspondence between the level of sex hormones in its bloodstream and an animal's sexual responsiveness. Aggressive behaviour also increases under the influence of male sex hormones and males become more responsive to stimuli which provoke attack.

Sometimes the threshold for response becomes so lowered that consummatory acts can be evoked by very minimal stimuli. Indeed Lorenz used the term 'vacuum activities' for those cases where the behaviour is produced in the absence of any external stimulus. Tinbergen[258] describes a number of such cases; for example, a starling is seen to go through the motions of catching and eating a fly when none is there. When deprived of nest material, a male Bengalese finch will go into its nest box and perform all the movements of carrying and placing material, although it has nothing in its beak. Clearly in a strict sense it is impossible ever to designate an activity as 'vacuum'; some external stimulus, no matter how minute, may be there. Nevertheless the term does draw attention to the extreme lowering of threshold which can occur.

To ascribe all such examples of changing responsiveness to a change in drive is no kind of explanation. The drawback, as mentioned earlier, is that by doing so we may give the impression that such changes are all produced in a similar fashion. We have little idea of the mechanism in most cases, but enough to indicate that it is not always the same. In Chapter 6 we shall consider some of the ways in which hormones affect behaviour. Sex hormones, for example, have often been described as increasing the sex drive and thereby changing responsiveness to sexual stimuli. We know that these hormones can act centrally on brain mechanisms, but they also affect responsiveness by peripheral action. Beach and Levinson[25] found that one action of testosterone in the male rat was to cause thinning of the epithelium of the glans penis. This increases the sensitivity of its tactile sense organs and may increase the rat's sexual responsiveness accordingly.

In most cases we assume that the incoming sensory information which results from a standard stimulus remains constant, but the brain's responsiveness alters. Only rarely do we have direct proof of this, but the elegant experiments of Dethier and his collaborators have shown that this does, in fact, hold during feeding in the blow-fly, *Phormia* (see Dethier[63]). When the chemoreceptors on its feet encounter sugar they can trigger off extension of the fly's proboscis onto the food where further sense organs determine whether it is sucked in. Direct recording from the sensory nerves originating in these sense organs shows that they always perform in the same way. They produce a burst of impulses when they first come into contact with sugar, but this decelerates back to the normal background level if the hair remains in the sugar—in physiological terms the sense organ adapts. The time it takes to adapt and the initial rate of firing depend on the sugar concentration. Low concentrations cause a very brief initial burst followed by rapid adaptation; high concentrations produce a prolonged, rapid burst and slow adaptation. These characteristics are the same whether the fly is fully fed or starving. but a hungry fly will respond to a more dilute sugar solution than one which has fed recently, and it will go on feeding for longer. The fly's 'acceptance threshold', as Dethier calls it, fluctuates with food deprivation although its sensory threshold does not.

The spontaneity of behaviour

Animals often start to behave in a particular way quite spontaneously. A dog wakes up on its own, stretches itself and then moves off in search of food. Of course the term 'spontaneous' can never be used very precisely. The dog, for example, may be woken up by hunger pangs, transmitted by nerves from its stomach. Nevertheless some behaviour does show a very striking degree of spontaneity and it is for this reason, more than any other, that the concept of an internal drive, which in some way 'energizes' the relevant behaviour mechanisms, seems indispensable to many behaviour workers.

In spring, a male chaffinch alone in an aviary, will begin to sing, and will continue singing for some minutes at a regular rate of some three songs per minute. It is very difficult to relate this song rhythm to any changes outside the bird's own nervous system. Nice[199] in her study of the American song sparrow observed that the tendency for males to sing gradually increased during early spring. Cold weather inhibits singing, and, over a number of years, she was able to plot how low the temperature had to be before singing was stopped on different dates. Fig. 4.1 shows a graph of her results; they strongly suggest that a spontaneous 'urge' to sing is increasing through January and February and accordingly cold has to be increasingly severe to suppress it.

At one time experimental psychologists viewed the concept of 'spontaneous behaviour' with scepticism. They favoured a more rigid stimulus-in/response-out type of organization, but this has proved quite inadequate to account for complex behaviour. In fact, neurophysiology has overtaken behaviour theory and shown that neurons themselves are spontaneously active and go through regular cycles of discharge in the absence of any

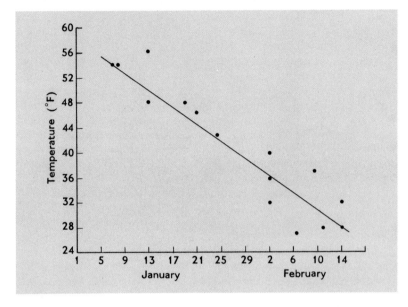

Fig. 4.1 The level to which temperature must fall in order to suppress singing by male song sparrows. As the season progresses the fall must be to lower and lower temperatures. (From Nice,[199] 1937, 1943, *Studies in the Life History of the Song Sparrow*. By permission of Dover Publications Inc., New York.)

input. Bullock[42] discusses this phenomenon and the ways in which such neurophysiological events may be related to behavioural ones. In Chapter 1, Roeder's[215] work on mantids was used to exemplify how the neurons of the thoracic and abdominal ganglia of this insect are spontaneously active unless held in check by inhibitory impulses from the head ganglia. In the latter's absence they initiate all the movements associated with copulatory behaviour, whether or not any external stimuli are provided.

There are no physiological objections to the suggestion that specific appetitive behaviour arises spontaneously as a result of increased activity in specific parts of the central nervous system. Such activity may indeed be independent of external stimuli in any direct sense, but may result from internal changes such as those which follow many hours of food or water deprivation.

If we call such spontaneous activity 'drive', we are using the term according to its original derivation, i.e. that which 'drives' or 'urges' the animal to do something. Drive in this sense may also be used to describe how, for example, a hungry animal will actually work harder to obtain food. Experiments which illustrate this aspect of drive will be described later when discussing measuring techniques.

The temporal grouping of activities

An animal's behaviour is normally well organized so that each pattern is brought into play when it is most effective. A male stickleback begins to build a nest by digging a pit in the sand; he then brings material to the pit and glues it together, pressing and moulding it into a solid structure, through which he eventually forces his way to make a tunnel which he then keeps open.

This sequence could be an example of a 'chain reaction', i.e. the end result of one pattern provides the stimulus for the next one to begin. Thus sand-digging produces a pit and an empty pit may be the stimulus for bringing material. However, we still have to explain why the stickleback only builds nests early in the breeding season and is quite unresponsive to nest-building stimuli at all other times. The thresholds for all these different but related behaviour patterns rise and fall together although we are not dealing with a rigidly fixed sequence which is predetermined.

Such examples have been explained by proposing that all the patterns share a common drive. This is taken to mean that they are all activated, at least potentially, and other factors, partly internal organization and partly external stimuli, determine which particular pattern appears at any given time.

Sometimes this type of hypothesis has been taken further and the linked series of behaviour patterns are ranged on an 'intensity scale'. This implies that the strength of the drive plays a large part in determining which pattern of the series is performed, because the patterns have successively higher thresholds. If drive is low only the pattern with the lowest threshold is activated; if drive increases this may be replaced by the next highest, and so on. Soon after settling on his territory, a male stickleback begins to dig the pit which will receive his nest. At first we may observe no more than his swimming low over an area of sandy bottom, later he dips his head repeatedly towards one part of the sand. Later still he touches the bottom and sucks sand into his mouth only to drop it immediately. Finally he sucks up sand, turns and carries it away before returning to collect more. The whole sequence, from the first *intention movements* of sand digging— as they are described—to the commencement of the pit in earnest, may take a few hours. Intensity changes of this type are commonly observed early in an animal's reproductive cycle at a time when we might expect drive to be increasing. To suggest that motivational changes underlie such behaviour changes is not to deny the part which external stimuli may play in the changeover from one pattern to the next. Stimuli resulting from the performance of an early pattern in the series may be one factor which causes the drive to increase and so activate the next pattern.

One example where the concept of an intensity scale seems to be useful comes from the work of Gardner[81] on the feeding behaviour of jumping spiders. These spiders do not spin webs, but stalk their prey by sight. When a hungry spider sees a fly it turns towards it ('orientation'); it then pursues the prey, running rapidly at first but slowing down to a cautious stalk as it gets closer ('pursuit'). When the spider is within jumping range it crouches, and after a brief pause jumps, so as to land on top of the fly.

Orientation, pursuit, crouching and jumping form the normal sequence of hunting. Spiders will catch and eat many flies in succession, and Gardner showed that as a hungry spider becomes full the sequence begins to be truncated. More and more sequences get no further than orientation, although if the spider begins pursuit he usually goes on to crouch and jump. Eventually the spider no longer even orientates to flies. There appear to be three stages on the intensity scale.

1. At the lowest level of hunger no response can be elicited.
2. At the next level the spider orientates towards prey.
3. At the highest level it goes on to pursue, crouch and jump.

These last three patterns are apparently closely linked and Gardner can find no evidence that crouching or jumping requires a higher drive than pursuit.

This is a good example of an adaptive sequence of behaviour patterns where an intensity scale can be used with some confidence and matched against another, quite independent, assessment of drive; i.e. how long it is since the spider last fed. Not all sequences can be treated in this way.

Hinde[109] has studied the nest-building behaviour of canaries. Building behaviour begins quite soon after a male and female are put together and gradually increases to a peak 3 or 4 days before the eggs are laid; female canaries do nearly all the building. The building behaviour can be classified into a number of categories such as gathering material, carrying it to the nest, placing it on the nest, and sitting, weaving and shaping the material into the nest. Each of these categories, in turn, consists of a series of more or less stereotyped patterns; sitting, for example, involves four or five different types of pattern.

Hinde used a standard observation period of 30 minutes and scored the number and duration of all these activities. There are some parallels with the spider hunting behaviour. Thus the more time spent in building, the more sequences, gathering, carrying and weaving, are completed. The transition from one activity to the next is only partly determined by the external stimuli from the nest, so there must be some internal factors which promote the change. Further, it is known that all the nest-building activities do share some common causal factors; oestrogen causes them all to increase in frequency, for example.

However, although it might seem reasonable to use total time spent building as a measure of a 'nest-building drive', this does not correlate with the proportion of time spent in the various constituent patterns. In other words, it is not possible to arrange the patterns on an intensity scale, nor is it possible to define the drive except as a tendency to nest-build.

To recapitulate this section, the concept of drive has been invoked to explain among other things, fluctuations in responsiveness to particular stimuli, the spontaneity of behaviour and the way in which a series of patterns appear grouped together. It is quite certain that no single entity with fixed characteristics can be involved. The concept must stand or fall on its usefulness; sometimes, as with feeding and drinking behaviour, 'drive' is a useful term whose physiological basis can be specified to some extent. In other cases, as with canary nest-building, it has no use save perhaps as a shorthand description of what we observe.

MODELS OF MOTIVATION

Having outlined some of the characteristics of goal-orientated behaviour with which the concept of drive has been associated, we should now turn to consider briefly whether we can offer any explanation for them in behavioural terms.

Our discussion will centre around two of the many 'models' of how behaviour is controlled. The purpose of a behaviour model is to devise a system of hypothetical components to which particular properties are given and which are so connected together that their 'behaviour' reproduces that which we observe. This is not such a useless operation as it may seem, because good models help to organize our thinking and suggest experiments which can test their own adequacy. If we find that one model consistently explains an animal's behaviour under a wide range of conditions, it may tell us a great deal about the *principles* upon which the nervous system is operating. It can tell us little or nothing about the *means* of operation. The nervous system uses large numbers of interconnected neurons for its operation and the interpretation of how the principles are put into practice remains a neurophysiological problem. A model can use components whose properties are borrowed from telephone exchanges, hydraulic systems or electronic components, so long as they can reproduce the result we are trying to interpret. Hinde[112] and the opening chapters of Deutsch's[65] book include excellent discussions of the principles and uses of behaviour models.

Sometimes ethologists have interpreted their observations with reference to a particular behaviour model which, in its final form (illustrated

Fig. 4.2 Lorenz's 'psycho-hydraulic' model of behaviour; see explanation in text. (From Lorenz,[157] 1950, *Symp. Soc. Exp. Biol.*, **4**, 256.)

in Fig. 4.2), we owe to Lorenz[157]. He uses components which are borrowed from a hydraulic system, and this model is often called a 'psycho-hydraulic' one. Lorenz envisages that when motivation is increasing, as when an animal is deprived of food, there is an accumulation of 'action specific energy', i.e. energy which is earmarked for feeding alone and does not affect other types of behaviour. In the model this is represented by the gradual accumulation of water in a reservoir (R) supplied from a tap (T). Outflow from the reservoir represents the motor activity of behaviour, but this is normally held in check by a valve (V) held shut by a spring (S). There are two ways in which the valve can be opened. Weights on the scale pan (Sp.) can pull it open and these represent various strengths of stimuli. The gradually increasing pressure of water in the reservoir and the weights on the pan both act in the same direction—to open the valve. The higher the water level, the smaller the weight required and eventually the water pressure alone may push open the valve—a vacuum activity. Lorenz represents the different types of motor output by a graded trough (G and Tr.). If the valve is opened slightly a little water trickles through

Fig. 4.3 Deutsch's model of behaviour; see explanation in text. (From Deutsch,[65] 1960, *The Structural Basis of Behavior.* © by the University of Chicago Press, Chicago.)

which only reaches up to the first and lowest hole in the trough. This represents the motor activity with the lowest threshold—often some form of appetitive behaviour. As the valve is opened wider, the trough discharges through other holes, which represent activities with higher thresholds and higher up the intensity scale. Once the reservoir is empty, the behaviour can no longer be elicited no matter how strong the stimulus; Lorenz talks about the 'exhaustion' of a behaviour pattern with this model in mind.

It is obvious that this psycho-hydraulic model accounts satisfactorily for cyclical changes in responsiveness. Note how quiescence following the consummatory acts is dependent on the *performance* of these acts, because this is the only way in which the reservoir can be emptied.

Now let us look at another model to explain the same set of facts. This one (Fig. 4.3) is put forward by Deutsch[65]. It forms part of a more comprehensive behaviour model which covers learning; here we illustrate only the section relevant to motivation. Deutsch uses a series of hypothetical components called 'analyser', 'link', etc., whose operation he defines. We may not immediately 'see' how the model operates, as we can that of Lorenz, because it does not use 'working parts', but its operation is very simple. Some deficit or imbalance in the internal medium, shortage of blood glucose or water for example, is detected and excites a central structure or link. The persistence and strength of this excitation depends on the size of the deficit. The link, in turn, activates the motor system which produces behaviour. (This simplified portion of the model does not include any representation of external stimuli which release the behaviour or any way of 'grading' the behaviour in intensity, but these are easily developed.) As a result of its behaviour, some aspect of the animal's environment changes. It has eaten or drunk and its stomach is now full of food or water, for example. The change in the environment is signalled to an analyser component which switches off the activity of the link so that it is no longer responsive to excitation from the internal medium. This inhibition slowly decays, until the link is once more sensitive to excitation.

To compare the merits of these two models we must see how well each accounts for the experimental data of behaviour. The psycho-hydraulic model is the more detailed of the two and does fit the facts derived from simple observation extremely well. It has been criticized because it uses hydraulic components which, some critics say, are too far removed from the actual operation of the nervous system. This type of criticism is quite invalid. There are no reservoirs filling up with water in the nervous system, nor are there black boxes labelled 'Link'; it is only the operating principles that count and the crucial test is which model fits the facts best.

Janowitz and Grossman[131] operated on dogs so that the oesophagus opened by a fistula to the outside of the throat. This meant that when such a dog ate the food fell to the outside and did not enter the stomach ('sham-eating' as it is called). The stomach could be filled with food from the outside without the dog having eaten anything. By this means it is possible to dissociate the behavioural act of eating from the normal results of having eaten. Suppose a dog with an oesophageal fistula of this type is kept for hours without food, but then has food placed directly into its stomach so as to fill it. Will the dog behave as if it is hungry or not? The psycho-hydraulic model would predict that it *would* still eat food because the reservoir, having filled with water (action-specific feeding energy in this case), has not had its valve opened and the water released through the trough (feeding behaviour); the reservoir remains full. Deutsch's model predicts that the dog will not eat. The link has been excited but nothing is 'stored' in this system, and although no motor activity has been performed, as soon as the environment changes (stomach full, in this case) the analyser is activated and this switches off the link so that no further feeding activity occurs.

Janowitz and Grossman found that their fistulated dogs do not eat after their stomachs have filled and thus, in this respect, Deutsch's model is better than the psycho-hydraulic one.

It appears that stimuli resulting from a distended stomach are the most important factor in switching off eating behaviour. Receptors in the throat may play a part, but a hungry, fistulated dog will continue sham-eating long after it would normally have stopped, if none of the food reaches its stomach.

With drinking, it appears that the mouth and throat receptors are more important than stomach distension. Bellows[26] found that thirsty dogs sham-drink little more than they would normally drink, but of course, since they did not replace their water deficit, they soon began sham-drinking again. These observations taken on their own, appear to fit the psycho-hydraulic model fairly well, but dogs which had their stomachs pre-filled with water soon ceased to respond to it although they had not performed any drinking behaviour.

It might be argued that feeding and drinking behaviour, because of their close relationship to an animal's physiological state are not typical of other types of behaviour where the goal may be a more purely 'behavioural' one. Perhaps in sexual behaviour, for example, it *is* the performance of the consummatory act which reduces the drive. However, in one instance, where it has been possible to separate the performance of sexual behaviour from its normal sequel, this has not proved to be the case. In the male stickleback there is a sudden drop in the sexual tendency immediately following the act of fertilization. Tinbergen[258] describes the stickleback's sexual behaviour in more detail; here we need only note that just after a female has laid eggs in his nest the male enters it and fertilizes them. Sevenster-Bol[238] has shown that the performance of fertilization is not necessary for the reduction in sexual tendency; it is also reduced if a male is allowed only to approach his nest and perceive the newly-laid clutch of eggs. This result again would not be predicted by the psycho-hydraulic model, but would by that of Deutsch.

Most recent evidence confirms that the period of quiescence following goal-orientated behaviour is the result of a sensory feed-back from some aspect of the situation which normally signals 'goal achieved'. This makes good biological sense. It is maybe an hour or more before the circulating food reserves of a starved animal are replenished by digestion of the food it has eaten, but its stomach is filled within a few minutes. Feeding behaviour must be stopped long before the physiological goal is attained.

One reason why Deutsch's behaviour model is more satisfactory than the psycho-hydraulic one is because it incorporates such a feed-back from the environment via the analyser to the link which is thereby switched off. As we shall see in a later section of this chapter, we can now identify parts of the brain whose properties resemble those of the components in Deutsch's model.

Another fundamental objection to the psycho-hydraulic model is its use of stored energy to represent motivation. This is most unlikely to have any counterpart in the functioning of the nervous system. An energy model of this type inevitably represents the initiation and termination of behaviour as two aspects of the same thing—energy starts to flow and energy runs out and is exhausted. In fact, as we have just seen, what switches on appetitive behaviour and what switches off the consummatory act and produces quiescence, may be completely different processes. Hinde[112] discusses the drawbacks of energy models of motivation in more detail.

MEASURING MOTIVATION—FEEDING AND DRINKING

Any study of behaviour in which motivation may be a factor requires that we can measure changes in the strength of this variable. We may be interested in how conflicting motivations interact or what effect strength of motivation has upon rate of learning, for example.

The methods we choose to apply for measuring motivation will depend both on the type of motivation and on the type of animal. We may try to measure the persistence of appetitive behaviour, or the frequency with which a consummatory act is performed. It is usually impossible to measure directly the internal state of an animal and we have to be content with measuring a response of some kind. This will be the result of the interaction between a stimulus and some mechanism which controls the performance of the response and whose properties will vary with the internal state. In fact we can keep the stimulus constant, and for the most part this indirect kind of measure is perfectly adequate to record changes in the internal state.

Feeding is certainly the best example to use when discussing various measures of motivation. All animals require food at regular intervals and it is easy to control their motivation by depriving them of food for varying lengths of time. The following list gives a few of the possible measures of the feeding tendency. They are conceived with a mammal, such as the rat, in mind, but some could be adapted for other animals. All of them could equally well be used to measure thirst.

   1. AMOUNT OF FOOD EATEN.   This is basically a measure of the consummatory act. It is usually easier to weigh the amount of food eaten than to count the number of feeding movements, for example. Most animals, when presented with food *ad libitum* will eat their way through it at a fairly constant speed until satiated. One obvious disadvantage of this measure is

that it reduces hunger as it proceeds. This might be a nuisance if one wanted to study how a measured level of hunger affected some other aspect of behaviour.

2. HOW BITTER FOOD CAN BE MADE BEFORE IT IS REFUSED. This measure, in effect, attempts to block the consummatory act of eating and one tests how far the animal will persist in spite of this. Quinine is an intensely bitter substance to ourselves and other mammals appear to find it likewise. A rat is presented with a series of tiny food pellets or drops of condensed milk, each adulterated with quinine. The concentration is gradually raised and at a certain level the rat will test the food with its tongue but reject it as too bitter.

3. STRENGTH OF PULL TOWARDS FOOD. In this, and the next measure, we might be said to block the animal's appetitive behaviour. A rat is fitted with a harness which can be engaged against a spring balance. If food is placed in full view at the end of a runway it is possible to measure how hard the rat pulls to reach it. It might also be possible to measure the speed of its run towards food in a similar apparatus.

4. LEVEL OF ELECTRIC SHOCK ACCEPTED. Again food is placed in full view, but to reach it the rat has to cross an electrified grid which it has previously learnt will shock its feet. By varying the level of shock it is easy to measure how much the rat will accept in order to reach the food.

5. RATE OF BAR-PRESSING FOR A FOOD REWARD. This, too, is a measure of appetitive behaviour, but it uses a behaviour pattern which the experimenter has purposely 'built in' to the food-seeking behaviour for the sake of convenience. The 'Skinner box' has already been described briefly (p. 18 and Fig. 2.1). It is a useful apparatus for quantifying a response which an animal has learnt. A rat is put into the box when hungry and is taught that it receives a small pellet of food when it presses a bar which protrudes into the box. (In Chapter 8 we shall discuss this sort of learning more fully.) When it has thoroughly learnt this, the apparatus is so arranged that rewards do not follow every press but come at irregular intervals, averaging out at—say—one reward every 30 seconds. In psychological jargon this is called a 'variable interval reinforcement schedule', and it means that the rat never knows whether any particular bar press will give results. Somewhat surprisingly perhaps, they press the bar much more regularly with such a schedule than they do if every press gets its reward. Because the rate is so regular under these circumstances it is suitable as a measure of hunger; the rate can be tested after different lengths of food deprivation.

We have outlined a number of possible ways for measuring hunger—many more could be devised. Superficially they all appear to be measuring the same thing, but we can learn something of the nature of feeding motivation if we compare how these different measures change with length of food deprivation. There is no study which combines all the five measures we have discussed, but Miller[189] describes a number of experiments which show that three of them, amount eaten, quinine accepted and rate of bar pressing, do not all rise together.

Over the range from 0 to 54 hours of food deprivation the amount of quinine rats will accept steadily rises, so does their rate of bar pressing, but their food intake reaches a maximum after only 30 hours and actually declines slightly thereafter. Thus a rat goes on signalling its increasing hunger by accepting food that is more and more bitter although it eats less. Presumably its stomach cannot hold more than a certain amount and the rat stops eating when it feels full. However, certain rats will eat very large amounts of food, yet show few other signs of 'hunger'. These are animals in which gross overeating—a condition called 'hyperphagia'—has been produced by making a small lesion in part of the brain which controls normal 'satiation'. (This result is discussed more fully later in this chapter.) Hyperphagic rats eat enormous quantities of food and become very fat, but they are not 'hungry' in the same sense as food-deprived normal rats. They are greatly handicapped by their obesity in any physical tests, such as pulling towards food, but they can easily press the bar of a Skinner box. They will do this, but at a much lower rate than a normal rat deprived of food for the same time. Further, hyperphagic rats are very 'finicky' eaters. The least interference puts them off and, for example, they will not accept as much electric shock to reach food as will normal rats, nor will they accept as much quinine.

Now it might be argued that rats with brain damage are so abnormal that nothing can be deduced from their apparent anomalies. However it seems incontestable that what we commonly lump under the term 'hunger' may be a conglomerate of factors and brain damage elevates some whilst depressing others. Miller[188-9] suggests that feeding and other types of motivation should, ideally, be measured in several ways and that a single measure may in fact give a totally misleading answer.

He[188] quotes some experiments by Choy on drinking behaviour. Rats had a tube implanted through the body wall so that liquids could be put directly into their stomachs without their having to drink. Choy measured what happened to drinking in a rat previously given *ad lib.* water, when 5 ml. of concentrated salt solution was put into its stomach. He measured in three ways: (i) amount drunk, (ii) concentration of quinine accepted and (iii) rate of bar pressing. His results are shown in Fig. 4.4, and the interesting point is the way in which the measures change relative to one another in the period following the administration of the salt solution. If we took bar pressing alone as a measure we would conclude that thirst is not increased for 15 minutes after giving salt. (The difference between experimental animals given salt and controls who were not is not significant after this time.) Thereafter, we would conclude, thirst goes on increasing for at least 6 hours. But this is only part of the truth; after 15 minutes the rats clearly are very thirsty in one sense because they begin to drink lots of water even though they will not press a bar any faster to get it. However, the amount they drink levels off 3 hours after the salt is given even though bar pressing

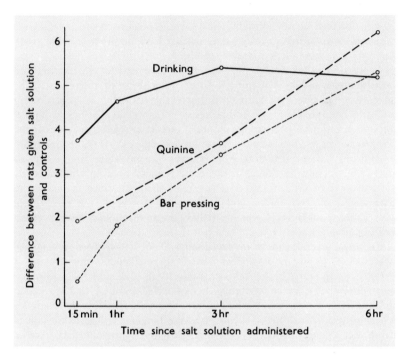

Fig. 4.4 How three different measures of thirst change in the period following placing 5 c.c. of strong salt solution directly into the stomach of a previously water-satiated rat. The units on the vertical axis are arbitrary; they are simply the difference between control and experimental rats on the various measures of thirst. (From Miller,[188] 1956, *Ann. N.Y. Acad. Sci.*, **65**, 318.)

continues to rise and so does their tolerance of quinine in the water. We need a combination of measures to get a reasonable picture of the effect of salt on thirst. As yet we do not know what lies behind the lack of correspondence between different measures, but this kind of data is an essential preliminary if we are to link physiology and behaviour and get at the real nature of motivation.

THE NATURE OF AGGRESSION

Ethologists are often interested in measuring the strength of sexual, parental or aggressive motivation. There is plenty of evidence to be discussed in the next chapter that, for example, the tendencies to attack and to escape may be aroused simultaneously and conflict with each other. Some measure of the strength of the tendencies is required, even if we cannot equate such tendencies with fluctuations in the internal state of the body as we can for feeding and drinking. Indeed, the need for good behavioural measures is more urgent because we usually have no other suitable yardstick, such as time of deprivation, to apply.

Aggression in particular presents many problems of description and measurement and it is important to examine some of them here. The aggressive behaviour of animals is currently attracting a great deal of attention from psychologists and psychiatrists as well as ethologists. This is because questions relating to the nature of human aggressiveness are of vital importance in modern society. How far has aggressiveness an inherited basis in man, descended to us from our ape-like ancestors? Is the expression of aggression inevitable or can it be reduced or even eliminated by certain types of upbringing? Discussion of such questions in relation to man would carry us far beyond the scope of this book, but the type of observations on animals which we have been discussing have a direct bearing upon them.

Recently several widely reviewed books[9B, 196B] have used ethological observations to justify a view of human aggressiveness which relates it very closely to man's biological past. Basically they all stem from Lorenz's view, developed in great detail in his book *On Aggression*,[160] that aggression in animals and men is the result of an inherited, spontaneous tendency whose properties are much the same as the biogenic drives to eat and to drink. Set against this is the view that, although aggression may have an inherited basis, there is nothing inevitable about it and its expression depends as much upon experience and external factors as the internal state of the animal. Amongst the chief exponents of this view are Barnett and Scott and essays by them and others reviewing Lorenz's book have been conveniently collected together by Montagu.[193]

What sort of evidence from animals has bearing on the problem? The question of how to define aggressive behaviour rises at the outset. Animals usually have recognizable and often stereotyped behaviour patterns for attack, but they may be used in several rather different contexts. In some cases there is a clear biological function for aggression as when animals fight to win a territory, to defend their young or to obtain food. Some species have a social organization based upon a stable hierarchy of dominance (see Chapter 10) and they may have to fight to maintain their status within such a group. Other contexts in which we observe aggression are less easy to explain. Pain arouses aggression; rats given small electric shocks attack a cage mate whom previously they ignored. Frustration of various types has a similar effect and one school of psychologists have suggested that all aggression is the result of frustration (see Miller[186A]). Clearly frustration of various kinds is one cause of aggressiveness in animals. A rat in a Skinner box will attack another rat tethered nearby if the bar he has learnt to press ceases to yield him the expected food rewards. In man too we are familiar with the effects of frustration—bad tempers abound in traffic jams. Nevertheless, it is extremely difficult to explain all animal aggression in terms of frustration, and a more direct, biological origin seems certain.

In predatory animals the patterns of intra-specific fighting may closely resemble those used in catching and killing prey. In spite of this, it is not usual to consider inter-specific predatory behaviour as aggressive and one assumes it is part of feeding behaviour. It seems best to consider aggressive behaviour patterns in the same way as one regards patterns of locomotion, as 'available' at the motor level of organization to more than one motivational system. If we exclude predation from our definition then the common link for the performance of aggressive patterns would appear to be a functional one; aggressive behaviour serves to displace another individual by causing injury or at least threatening to do so. (We shall discuss the specialized use of the term 'threat' in the next chapter and also consider there some of the physiological correlates of aggressive arousal.)

For man it is much more difficult to arrive at a satisfactory definition of aggression, because our behaviour takes such variable forms. In Carthy and Ebling's book[47] various authors propose that a whole range of behaviour from nail-biting through verbal insults to suicide are all manifestations of aggression. The aggressive behaviour of a street gang may involve actual violence which is related to their concept of 'territory' and therefore, at the functional level of analysis, is comparable to some animal aggression. Such behaviour will also involve the participants in the same kind of physiological arousal that we can measure in animals. Not all behaviour which is commonly called aggressive shares these features. In modern warfare an individual's act of pushing a button may lead to the destruction of other individuals at a great distance and of whom he has no direct knowledge. In a biological sense the button-pusher is not aggressively aroused and such behaviour defies any simple, biologically-based definition. Lorenz[160] and Tinbergen[263A] have drawn attention to the dreadful dilemma of modern man whose technology permits him to burn alive people whom he has never seen, and thus precludes any human contact which might inhibit his taking such an action.

From the question of defining aggression, we may return to examine some of the evidence which bears on the original problem of how far aggression can be compared with a biogenic drive such as feeding. Following Hinde,[112B] we can suggest several inter-related questions here, discussion of which helps to characterize aggressive behaviour more clearly.

Has aggression an inherited basis?

There is now a great deal of circumstantial evidence that, whatever their previous experience, many animals have a tendency to respond aggressively when first placed in certain situations or given certain stimuli. Often the situations and stimuli that evoke aggression relate to the setting up of a territory at the beginning of the breeding season and we shall be discussing territorial behaviour in more detail in the next chapter. Here we should note that aggressive responses are far commoner amongst males (and as mentioned earlier the male hormone testosterone tends to increase aggressiveness) and that the stimuli to which they respond are often those provided by a rival male. Thus Cullen[57A] has shown that sticklebacks reared in complete isolation from the egg, set up territories and attacked rival males in the typical manner. Isolation experiments have certain limitations, as we discussed in Chapter 2, but experiments of this type and the rôle of aggression in the organization of territorial behaviour in such a wide variety of animals, certainly suggest that there is an important genetic component in the development of aggressive behaviour.

However, even if we accept that the potential to perform aggressive behaviour has an inherited basis, this is far from establishing aggression as a biogenic drive. On page 62 we mentioned some of the difficulties in identifying either regular fluctuations in aggressiveness or the phases of appetitive behaviour, consummatory act and quiescence, such as characterize feeding and drinking. This leads to a second question.

Is there evidence for aggressive appetitive behaviour?

In other words, do animals actively move about looking for fights? We are all familiar with the sight of an aroused dog or cat pursuing another which it has just put to flight and clearly making no attempt to break off the encounter. But will a hitherto restful cat get up and move off to seek an opponent when none is visible? Birds are sometimes said to be 'patrolling their territories' and they will attack any intruder they come across, but it is impossible to be certain that it was aggressive motivation that initiated their patrolling. We would have to demonstrate a lowered threshold for aggressive responses just as the bird set off and before it had seen a rival and such measurements are difficult to make.

However, approaching the question from a slightly different angle, there can be no doubt that under some conditions aggressive behaviour is reinforcing, i.e. an aroused animal will perform behaviour giving it the opportunity to respond aggressively. Perhaps the most elegant examples of this come from the work of Thompson with game fowl[252C] (see p. 98) and Siamese fighting fish[252B] (both, significantly, animals which have been selectively bred for their aggressiveness). Fighting fish males will show the full aggressive display with raised gill covers and extended fins towards a mirror image or a model, and they will also attack and bite such stimuli when first shown. Thompson kept males alone in tanks where it was possible to make the male's image appear rapidly simply by switching off an outside lamp so that one wall of the tank functioned as a mirror. He then conditioned the fish in a manner exactly analogous to the rats in the Skinner box described on page 19. The latter had to press a bar to get a food pellet and soon learnt to do so. Thompson's fish had a ring suspended in the water and if the fish swam through it the lamp was automatically switched off for a few seconds. An image then appeared on the glass, to which the fish displayed aggressively. Under these conditions, the males learned to swim through the ring and repeatedly did so, some hundreds of times each day. In other words, given the opportunity to control its own situation the males chose one where they would be behaving aggressively for much of the time. In a similar fashion game fowl cocks would learn a response which gave them the opportunity to perform their aggressive display. It might be possible to argue that the presence of the ring in the tank became an aggressive stimulus and kept the fish aroused. Thompson has shown that artificial objects associated with the appearance of a rival male do themselves come to elicit attack. Even so, the conclusion that aggressive behaviour is 'rewarding' to the fish once aroused seems inescapable and does, in this respect, justify a comparison between the tendency to fight and that to feed or to drink.

Is there a phase of quiescence following the performance of aggressive behaviour?

The often stereotyped patterns of aggressive displays and attacks would seem to be the equivalent of the consummatory acts directed towards the goal object—e.g. the rival. However it is difficult to detect any regular phase of quiescence following their performance. With other types of behaviour, even feeding and drinking, it is common to find that there is a short period of arousal and warm-up after performance begins (see p. 7). Thus the intensity of the behaviour may transitorily increase before feedback signalling 'goal-achieved' begins to take effect and leads to quiescence. With aggressive behaviour such arousal is intense and once attack begins the animal's threshold for further attacks often falls considerably.

Some examples of this phenomenon come from the work of a group of

ethologists who have made an extensive study of the reproductive behaviour of the stickleback. This has involved them in developing measures
for the strength of sexual and parental tendencies in the male as well as
measuring aggressiveness. The type of measure used has to be suited to the
behaviour and it is rarely possible to use such a selection of measures as we
listed for feeding. For aggression, Sevenster,[237], Sevenster-Bol[238] and
Wilz[283C] have counted the number of bites delivered to a standard stimulus
during a standard time period (often 1 minute but sometimes as long as 10
minutes). The stimulus used was a live male in nuptial colouring showing a
red throat (see Fig. 3.2) confined in a narrow glass tube which was suspended at a fixed distance from the nest of the male being measured.

The frequency of bites may be taken as a direct measure of the consummatory act of aggression, and simple frequency measures of this type take
no account of the intensity of performance—equivalent perhaps to the
effort a rat expends pulling against a spring balance to reach food. Sevenster admits that bites do vary in their intensity, and the male does perform
other types of behaviour, e.g. threatening (see Chapter 5), towards the test
stimulus. However the lack of an intensity measure is unlikely to upset the
validity of one using frequency alone, because there is probably a positive
correlation between intensity and frequency. It is unlikely that some highly
aggressive male stickleback would content itself with a few, but very powerful bites to a rival!

From our point of view here, one significant feature of such aggression
tests is Sevenster's and Wilz's finding that a male's tendency to bite is
higher at the end of a 10-minute test than at the beginning. We would
certainly not expect this were an animal given adlib food for 10 minutes,
and in this respect the striking arousal of aggressiveness, once attack has
begun, is in strong contrast to the biogenic drives. Thus a quiescent stage
following the consummatory act is not at all evident. In the continued
presence of an aggressive stimulus attacks eventually cease, but a change of
stimulus will often start them again, just as a change of stimulus immediately restores a bull which is apparently exhausted sexually. It is
probably most appropriate to regard this waning of response as habituation
(see p. 177) rather than as quiescence resulting from a fall in motivation.

Using the same argument, the recovery of aggressive responsiveness
after waning, which may take days in Siamese fighting fish which have
had a long initial exposure to a mirror image,[54A] is just as likely to be due
to the waning of habituation to the stimulus as to an increase in motivation.
There is little point in overemphasizing one explanation or the other at
present because, in any case, the animal data is not consistent. Rather in
contrast to the example just given, Heiligenberg[102] found that in the
cichlid fish *Pelmatochromis* the tendency to attack *decreased* over a few days
in the absence of any arousal.

To summarize this section, the data from animals indicates that, unlike the biogenic drives to feed and to drink, the tendency to attack often *increases* as a result of the performance of aggressive behaviour. Undoubtedly aggressiveness fluctuates, but there is no critical evidence that it increases steadily in the absence of any external stimulus or opportunity to attack.

Such conclusions are of considerable theoretical and practical importance in the controversy about the nature of aggression, because of Lorenz's view of the ways in which both human and animal aggression can be controlled. Discussion of this can be directed to a fourth question.

Is aggression inevitable?

The reason for phrasing the question in this way relates to Lorenz's application of his psycho-hydraulic model to aggression. It will be recalled that with this model the only way to reduce drive was to perform behaviour and in the absence of an 'outlet', drive accumulated and became stronger.

If human beings have an inherited aggressive tendency which operates in this way then clearly we shall have to accept that it will be impossible to prevent all manifestations of aggression. The best policy would seem to be one of encouraging the sublimation or redirection of aggressive tendencies into outlets less harmful than those of physical conflict. This view is, in principle, that held by Lorenz, Ardrey and Morris.

There are a number of experiments which are often quoted as showing the inevitability with which aggression expresses itself, but they are all open to other interpretations. Animals reared in total isolation are sometimes highly aggressive. This has been clearly shown in mice and in jungle fowl, where in the latter species, Kruijt[142A] found that after months of isolation the birds would attack feathers and have prolonged, circling fights with their own tails. However the behaviour of isolates is profoundly altered in a number of ways and one is not justified in interpreting these observations as the sole result of an accumulating aggressive drive. Thus isolated rodents are generally highly excitable and are known to undergo hormonal changes—probably resulting from the stress of their situation— such that the gonads of males may be stimulated to secrete more testosterone. This, by itself, might cause an increasing tendency to attack.

Lorenz[160] described observations on a highly aggressive cichlid fish, *Etroplus maculatus*. Fish breeders have found by trial and error that in order to get a pair of *Etroplus* to breed successfully, it is necessary to have one or two other non-breeding males in the tank who serve as 'whipping boys'. The breeding male attacks them from time to time, but little aggression is seen within the pair. If a pair of *Etroplus* are kept alone, they rarely breed successfully because the male constantly attacks the female. These observations have been confirmed under well-controlled conditions by Rasa[210A] and Fig. 4.5 illustrates her results. There is a huge increase in attacks on the female if the male has no other fish to attack. Lorenz[160] and Eibl-Eibesfeldt[69A] interpret this result in terms of accumulating aggressiveness in the male which must find an outlet. However the lowering of the attack threshold does not necessarily imply accumulation. As we shall be discussing more fully in the next chapter, in territorial animals the presence of a female commonly arouses aggression in males in addition to

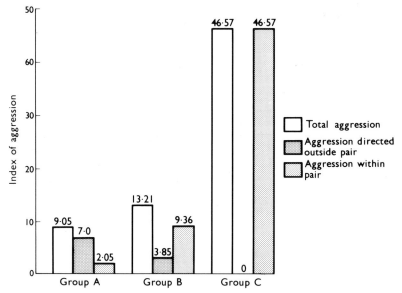

Fig 4.5   Rasa's data on aggressive behaviour in the cichlid fish, *Etroplus maculatus*. The 'index of aggression' represents the average frequency of chasing attacks made by both fish of a pair within five minute periods. In group A pairs were kept in communal tanks with other fish ; in group B pairs were kept separately but could see other fish through a glass screen ; in group C pairs were completely isolated. Aggression is classified into two categories, that directed towards other fish and that between the members of a pair—the only category in group C. (From Rasa,[210A] 1969. *Z. Tierpsychol.*, **26**, 846.)

sexual responses. Consequently the male *Etroplus* are likely to be aroused under all Rasa's experimental conditions. Their aggression is, to some extent, inhibited because the female also arouses sexual tendencies and, where other fish are present, most of the attacks are 'redirected' towards them. Redirection is familiar in the aggressive behaviour of animals and men when a stimulus which arouses aggression is also, for some reason, not available for attack. The boss reprimands his clerk, who in turn cuffs the office boy.

From Rasa's experiments we may note that the actual incidence of attacks rose enormously in the isolated pair condition. She suggests that this is because the male *Etroplus*, like the stickleback, is most aggressive close to his nest. This is where the female tends to remain, even when attacked and consequently we may expect the male to be maximally aroused. Further in the normal condition, the distant attacks he makes upon other fish will lead him away from the nest. Then, because the other fish retreat out of sight as far as they can, the *Etroplus* male's aggressiveness may have more chance to subside before the next stimulus appears. If we are to examine critically the concepts of accumulation and inevitability we need much more information on the behaviour of fish in which arousal is kept to an absolute minimum for varying periods of time. As yet we lack the necessary data.

Lorenz's concepts have been criticized by several of the mammalian workers because he has tended to underestimate developmental factors affecting aggressive motivation. Scott[234A] gives a very clear survey of numerous experiments with rodents which show how levels of aggressiveness can be changed in a most dramatic fashion by regulating early experience. It is relatively easy to train one mouse always to attack a strange animal, whilst another of the same strain can be trained to remain completely placid. In view of such undoubted effects and considering how equivocal is the other animal data there seems no reason why we should accept human aggressiveness as being inevitable. We certainly have to accept that man has a potential—probably inherited—for aggressiveness, but the long period of childhood development and the powerful influences that both parents and society can bring to bear on the individual offer us a solution.

THE PHYSIOLOGICAL BASIS OF MOTIVATION

Continued analysis of this behavioural type is essential, but it is also important to examine motivation at the physiological level and try to link up behaviour with events in the nervous system. Some of the most promising bridges between neurophysiology and behaviour have developed from the work of physiological psychologists—most of them Americans—on motivational problems. Grossman[88B] provides a good introduction to the whole field of physiological psychology. We must now turn to consider some of this work and begin with some description of one area of the vertebrate brain which has proved to be of major importance in the control of motivation—the hypothalamus.

The hypothalamus

This relatively minute volume of brain tissue—in the human brain it is smaller than the last joint of the little finger—is of primary importance in a whole host of reactions. There is an excellent general account of its physio-logy in Walsh[272] who says of the hypothalamus, '. . . this small centre plays a dominant rôle in determining the use that is made of the resources of the body. . . . It is difficult, indeed, to think of any function of the body that is not dependent, directly or indirectly, upon the hypothalamus.'

A little neuro-anatomy is needed at this point; the reader is referred to a clear, concise account in Romer.[216]

The brain of all vertebrates is constructed on the same basic plan, and at an early stage in embryology consists of three swellings at the anterior end of the spinal cord. These are called the prosencephalon, mesencephalon and rhombencephalon or, more simply, the fore-, mid- and hind-brain. Primitively these swellings arose to cope with the increased

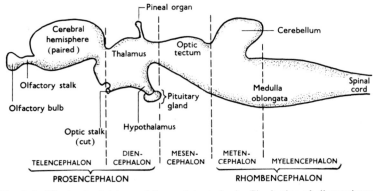

Fig. 4.6 The basic divisions of the vertebrate brain. The brains of all vertebrates pass through a stage rather like this during development, but in mammals and birds in particular, the adult brain is dominated by the enormous growth of the cerebral hemispheres and cerebellum. These come to overlie all the rest and obscure the original lay out; see also Fig. 9.1. (Modified from Romer,[216] 1962, *The Verte-brate Body*, 3rd edn. W. B. Saunders, Philadelphia and London.)

amount of sensory information which flowed into the central nervous system from the sense organs of the head. The fore-brain originally dealt with olfaction, the mid-brain vision and the hind-brain balance and hear-ing. In most living vertebrates their original functions have become greatly extended and complicated, but still the appropriate sensory data are led first to these regions, even if subsequently they are passed on elsewhere.

The fore-brain is easily sub-divided into two portions, as shown in Fig. 4.6. The anterior portion has arising from its roof the cerebral hemispheres, which primitively were olfactory areas but have now come to dominate the whole nervous system in mammals. The posterior portion of the fore-brain —called the diencephalon—has on its dorsal surface the pineal organ. This was once associated with a light receptor or pineal eye which can still be seen in some living reptiles. The side walls of the diencephalon are thick and form the thalamus, an important 'staging place' in the brain where fibre tracts link up with one another in numerous 'nuclei' or clusters of neuron cell bodies. On the floor of the diencephalon, below the thalamus as its name implies, is the hypothalamus.

There are nuclei in the hypothalamus, but they are not so well defined as in the thalamus above. However, there are several well-marked fibre tracts entering and leaving, and these put the hypothalamus into connection with the cerebral hemispheres and also with more posterior parts of the brain. From the behavioural point of view, one of the most significant structural features of the hypothalamus is its intimate connection with the pituitary gland. This endocrine gland, which controls the whole hormonal system of the body (see Chapter 6), develops from the fusion of a down-growth of the embryonic hypothalamus with an upgrowth from the roof of the embryonic mouth cavity. The pituitary stalk, which joins it to the hypothalamus, contains both nerves and blood vessels. The hypothalamus itself has a very rich blood supply and some of its cells are even penetrated by capillaries.

From its connections with other parts of the brain, its rich blood supply and its links with the pituitary gland, the hypothalamus is well adapted both to measure changes in the metabolism of the body and to set in motion activities which will rectify them. It is, in other words, well suited to serve as part of a homeostatic control system and there is plenty of physiological evidence that it does so. For example, the control of body temperature is one of the most delicate homeostatic systems in a mammal or a bird. There are areas of the hypothalamus which are highly sensitive to changes in the temperature of the blood. If these areas are heated artificially by implanted wires, the animal starts sweating and panting. Sweating is controlled peripherally by the autonomic nervous system and the hypothalamus can initiate its activity. The reverse effect is produced when the temperature-sensitive areas are cooled; now the animal shivers—another autonomic response (see Walsh[272]).

All this may appear to be pure physiology, having little to do with the study of behaviour. But homeostasis is one of those topics where the boundaries between traditional fields of study break down. It is not helpful to distinguish between physiology and behaviour in the matter of tempera-ture control. If a rat is briefly cooled in the manner described, shivering is started to generate some heat. We might classify this as a reflex activity and consign it to the realms of physiology, as described in Chapter 1. However, if the rat is cooled for a longer period shivering alone is inadequate and, if given the material, the rat begins to build a nest or to enlarge the one it already has, in order to insulate itself. The reflex response is now supported by a complex behavioural one. Both are initiated by the hypothalamus and both form part of the rat's homeostatic system, although nest building will involve a more elaborate neural mechanism than does shivering.

The hypothalamus and motivation

It is from studies of the rôle of the hypothalamus that some of the most important links between brain and behaviour have developed. Modern physiological techniques allow parts of the brain to be explored with electrodes, controlled injection of chemicals or by the destruction of very small selected areas.

As a typical example of the way such studies reveal the underlying physiology of motivation we may consider the rôle of the hypothalamus in thirst. There are cells in its lateral areas which respond to increased concentration of the circulating body fluids. They can set in action two compensating systems. The first, acting via the links with the pituitary gland, causes the secretion of antidiuretic hormone (ADH) which increases the resorption of water by the kidneys. The second system causes the animal to seek out and drink water. If the link between the hypothalamus and the posterior lobe of the pituitary gland is damaged, ADH may never be secreted. In such a situation the kidneys continue to excrete copious quantities of urine and to compensate the animal drinks large quantities of water—a condition known as *diabetes insipidus*.

Normally the amount of water taken in is exactly adjusted to the animal's needs, but if the lateral hypothalamic detector area is artificially stimulated either electrically or by injecting hypertonic saline, drinking is greatly increased. Andersson[5,6] prepared a number of goats with fine hollow needles penetrating into the hypothalamus. He allowed them to drink as much water as they would take and then injected concentrated salt solution. There was little effect unless the needle tip was in the lateral hypothalamus. In this region the salt caused the goats to drink frantically within a minute or two of injection. These animals, previously satiated with water, would drink salt and bitter solutions which they would not normally touch even if extremely thirsty. This condition is not comparable to *diabetes insipidus*, because the goats are not compensating for water lost through the kidneys. They are drinking water which is surplus to their physiological needs.

Andersson[6A] could produce the same effect if he stimulated the lateral hypothalamus electrically and such drinking behaviour is often described as 'stimulus-bound', because it continues only whilst the stimulus (saline or electrical) is actually being applied. Stimulus-bound drinking has also been produced in the rat and the importance of the lateral hypothalamus for drinking is further emphasized by the behaviour of rats with damage to this part of the brain.[252] Such animals may stop drinking altogether. They will not drink even though in the last stages of dehydration and will eventually die in the presence of water, unless this is given artificially through a tube into the stomach. It is fascinating that a rat with a lateral hypothalamic lesion is not just disinterested in water, it becomes actively averse to it. If water is placed in its mouth, the rat will not swallow but allows the water to run out, making tongue and lip movements identical to those made by a normal rat towards an intensely bitter liquid.

These effects of stimulation and ablation might lead us to conclude that the lateral hypothalamus is responsible for the state we observe behaviourally as a tendency to drink. How far is this conclusion justified? Before discussing this question we can add comparable results which also implicate the hypothalamus in the control of feeding.

Mammals and birds normally keep their weight very constant and adjust the amount they eat accordingly. If rats are given a super-rich diet they eat less; if their food is mixed with non-nutritive cellulose they eat more. We have already mentioned that rats with damage to central areas of their hypothalamus (the ventromedial nucleus, in fact) lose this sensitive control of their eating. They become grotesquely fat and continue eating far too much. Electrical stimulation in this area depresses feeding, so that the ventromedial nucleus has been called a 'satiety centre'; it measures when the animal has eaten enough and inhibits further eating. (The term 'centre' here means a group of nerve cells with a common organizing function; see p. 91 for a further discussion of this term.)

A complementary centre which promotes feeding is situated in the lateral hypothalamus, closely associated with the drinking area. Its cells are responsive to changes in the bloodstream during food deprivation. The exact nature of these changes is still disputed, but perhaps involves the level of glucose in the blood. In rats, experiments similar to those described for thirst have shown that this 'feeding centre' does initiate the appetitive behaviour of eating; stimulus-bound feeding can be elicited by implanted electrodes and damage to the area can lead to starvation in the presence of food. The feeding and drinking centres are closely linked anatomically and they interact with each other in a complex fashion. For details of the behaviour associated with this interaction the reader is referred to the admirably clear account by Teitelbaum and Epstein.[252]

How far are we observing the operation of the 'normal' drinking and feeding systems during stimulus-bound behaviour—remembering that this term does not refer to 'normal' stimuli but only to artificial stimuli applied chemically or electrically within the brain itself? It would be important to see whether an animal shows normal appetitive behaviour when being stimulated, because it might be argued that their feeding or drinking was merely a reflex response to the stimulation of neural pathways controlling the motor patterns concerned. It is possible to get fairly well co-ordinated lip and tongue movements by stimulating other areas of the brain—parts of the motor cortex of the cerebral hemispheres, for example.

Some of Andersson's observations appear to rule out this explanation because, when stimulated, his goats showed all the signs of normal thirst. They walked over to the corner of their pen and searched around for the water bowl, i.e. they showed normal appetitive behaviour and not just 'forced drinking'. Again with feeding in rats, Coons et al.[54B] have shown that stimulus bound feeders will learn a new response (bar pressing) in order to acquire food, and that this response is transferred and used when subsequently the same animals are made normally hungry.

Such results are very convincing, but some doubts still remain and have been emphasized recently by Valenstein[266C] and others. There are a number of experiments which have shown that stimulus-bound feeders and drinkers are much more easily dissuaded from their goal than normally motivated animals. At least this is the case in rats, where slight adulteration of food or water with quinine is usually enough to stop them responding. (We have already mentioned how easily hyperphagic rats are dissuaded by quinine.) Further, the rate at which rats drink by lapping from a tube is normally very constant. Normal water deprivation simply leads to lengthening their bouts of drinking but does not affect their lapping speed. However White et al.[281B] report that stimulus-bound drinkers do change their rate of lapping with the intensity of electrical stimulation, which certainly suggests that the motor-organization centres for drinking are being affected.

We shall need more facts to be able to resolve this question and, even from the evidence we have at present, we must not expect the same details to apply to all vertebrates or even to all mammals. Nevertheless there can be little doubt that the basic function of the hypothalamus as a detector of physiological imbalance remains constant. Its cells will meter the temperature, food, water and hormonal concentrations of the bloodstream. As a result of any particular imbalance, the detector sites will initiate both physiological and behavioural action, as we have outlined earlier in this section.

The behavioural problems then become more obscure, because whilst we may accept that the hypothalamus is essential for the initiation of motivational states, we cannot claim that it is the only essential for their control. Clearly this control involves many other areas of the brain also. Grossman[88C] reviews the extensive evidence showing that other parts of the forebrain are involved in feeding, drinking, sexual and aggressive behaviour. Many experiments have involved making lesions and observing their behavioural effects. It has been found that damage to most areas of the cerebral hemispheres does not affect an animal's motivation in a specific way, but there are some significant exceptions. Many of these concern lesions in the extreme frontal parts of the hemispheres and also in a complex series of fibre tracts called collectively the rhinencephalon or limbic system. These are situated near the base of the cerebral hemispheres and have links with the hypothalamus. Parts of the limbic system used primitively to be connected with the sense of smell, but in higher mammals it is clearly involved in more general aspects of their behaviour and appears to modify the activity of hypothalamic centres.

Lesions in the frontal lobes of the hemispheres and in the limbic system have been found to affect the sexual and feeding tendencies of dogs, cats, and other mammals and sometimes in a highly dramatic manner. For instance, dogs with damage in parts of the frontal lobe show ravenous eating, devour inedible materials and become grossly obese. Cats with lesions in the region of the amygdala (a part of the limbic system) became hypersexual, mounting and attempting to copulate with a wide variety of animals and inanimate objects (see Grossman[88B] for further details).

There is some evidence that a loop of neural tracts in the forebrain are involved in the control of feeding and drinking. Tracts from which such behaviour can be elicited on stimulation leave the hypothalamus, proceed through various limbic structures and can then be traced back to the hypothalamus again. Of course such a circuit is not closed—it will have outside connections along its length—but perhaps continuing activity within such a loop is one basis for a motivational state.

Summarizing the evidence on control centres

In conclusion we may recapitulate the evidence on the neurophysiological basis of motivation and consider, in particular, whether specific motivational states can be initiated by the activity of discrete centres in the hypothalamus.

The concept of a 'centre' is rather imprecise but it is usually taken to mean a relatively small group of nerve cells with a common function. Various techniques have been used to explore the organization of the brain and, as we have seen, one of the commonest techniques used by physiological psychologists is to make lesions and study their effects on behaviour. There are dangers in trying to locate centres by making lesions because the brain is an amazingly complex structure made up of minute units. Burning holes in it or cutting pieces out is inevitably a rather crude technique. In addition to removing many cell bodies, nearly always some fibre tracts are damaged which lead from or through the area of the lesion. Other areas, supplied by such tracts but not necessarily close to the site of the lesion, may be affected by it and this may make interpretation of the results difficult. For instance, Reynolds[212] has suggested that the ventromedial nucleus of the hypothalamus is not a 'satiety centre'. He considers that the lesions which are made to ablate it result in constant 'irritation' of tracts linking this nucleus to the lateral hypothalamic 'feeding centre', and it is stimulation of this latter area which causes hyperphagia.

Gregory[87] uses a vivid analogy to emphasize the problems of interpreting the results of lesions. Discussing how far comparing the brain with
various types of machine is helpful for understanding its operation, he
says, 'Thus the removal of any of several widely spaced resistors may cause
a radio set to emit howls, but it does not follow that howls are immediately
associated with these resistors, or indeed that the causal relation is anything but the most indirect. In particular, we should not say that the
function of the resistors in the normal circuit is to inhibit howling. Neuro-
physiologists, when faced with a comparable situation, have postulated
"suppressor regions".'

The evidence gained from lesions needs support from other types of
experiment if we are to identify control centres with any confidence.
Suppose we are trying to demonstrate that a certain area of the brain is
responsible for starting one type of specific behaviour, related to a particular goal. Ideally we might hope to use the techniques set out in Table 4.1
and record the equivalent results. These are a somewhat hypothetical
series of experiments and the postulated results are over-simplified, but
there is information comparable to this for the lateral hypothalamic
'feeding centre' and for the corresponding ventromedial 'satiety centre'*
Similarly there is good evidence that a centre in the posterior hypothalamus picks up female sex hormone from the blood and controls the onset
of oestrus behaviour. We shall discuss this and related results more fully
in Chapter 6. Aggression and fear centres are much more problematical.
Attack and 'rage' responses in cats, for example, can be obtained by
stimulating many different areas of the hypothalamus and limbic system.
It is necessary to be particularly careful with interpretation here because all
sorts of painful stimuli can provoke rage and the electrodes may sometimes be stimulating pain pathways. Nevertheless, whilst individual
elements of the cat's 'rage' response such as spitting, hair erection and
arching of the back can be elicited from other areas of the brain, the
hypothalamus must be intact if these are to be integrated into the full
pattern.

Recently Valenstein[266C] and his collaborators have questioned the
specificity of these proposed centres. In part, their approach takes us back
to the discussion of general versus specific drives with which this chapter
opened. They have found that *different* behaviour—feeding or drinking—
can be elicited from the same electrode in the lateral hypothalamus,
depending on the experience of the animal in the situation. They suggest
that stimulation arouses the animal in a general fashion and facilitates any
responses which—to use their term—are 'prepotent'. This prepotency will

* It is interesting to note that these centres thus appear to have properties
analogous to the 'link' and 'analyser' respectively of Deutsch's behaviour model
(see p. 71 and Fig. 4.3).

be affected by the internal state of the animal and the external stimuli. The exact experimental procedures and the evidence they provide are too complex to discuss here. A very good picture of them is gained by reading the dialogue between Wise[283D, 283E] and Valenstein et al.[266A, 266B] in Science.

Wise argues the case for specific drive centres, although he does not deny that the feeding and drinking centres are very close and that both are liable to be affected by diffusing current from the same electrode. Such diffusion is always a potential hazard of brain stimulation work and for this reason Miller,[190] Grossman[88A] and others have advocated the use of chemical stimulation. The central nervous system is highly diverse biochemically and this could make chemical stimulation much more selective

Table 4.1 Possible techniques to be employed in exploring a postulated 'centre' controlling a particular type of specific behaviour.

Technique	Expected Result
1. Ablate area of brain	Specific behaviour not shown even when conditions are optimum for it.
2. Stimulate area $\begin{cases} \text{electrically} \\ \text{chemically} \end{cases}$	Specific behaviour shown even when completely inappropriate.
3. Depress area chemically	As for (1)
4. Record normal electrical activity from area in conscious, freely-moving animal.	Activity high when specific behaviour is being performed. Activity low when not being performed.

than making lesions or stimulating electrically. Both electric charge and chemicals will diffuse out from their points of origin and affect several adjacent 'circuits' but, whereas all neurons respond to electrical stimulation, each circuit may respond only to a few specific chemicals. This means that successive injections of a series of different chemicals may affect 'circuits' one at a time and such results are more easily interpreted. Using such a technique Grossman succeeded in selectively increasing either feeding or drinking by specific chemicals injected through the same needle into the lateral hypothalamus. Since these effects were so distinct, it seems unlikely that the lateral hypothalamus is as undifferentiated as Valenstein implies.

Whatever view is taken in this controversy, most workers would agree that it is impossible to pin down drive centres to a small, unique group of neurons. Indeed it is probably misleading to think of centres in these terms, because the brain probably functions using larger and more diffuse groupings of its units. The feeding and drinking centres are probably the nearest approach to the ideal, but even here exact localization is not possible.

The elegant work of Teitelbaum and Epstein[252] shows that there is considerable plasticity in the cells of the lateral hypothalamus. Rats which have stopped eating altogether following a lesion in this area can be coaxed back into normal feeding again by starting with an attractive liquid diet. They regain first the desire to eat and later the ability to regulate their food intake. This must mean that neurons outside the original feeding centre have come to take over its function. The ability to compensate for the effects of damage is widespread in the brain; it rules out too rigid a localization of function. This conclusion is reinforced by the awkward fact, which we have already mentioned in respect to Valenstein's work, that stimulation at the same site on different occasions may not always produce the same results. Von Holst and von Saint-Paul[118] describe numerous examples of this from their work on brain stimulation in chickens. Often they could correlate the changed effects with other spontaneous changes in the birds' responsiveness to particular stimuli. The activity of one control system will affect that of others and may change the threshold at the site of electrical stimulation, thus determining into what channels stimulation spreads.

We must not expect that artificial stimulation—nor the natural external stimulus situation—will arouse only one, isolated motivational system. In the next chapter we must consider in more detail what happens behaviourally when two different systems are aroused simultaneously.

5

Conflict Behaviour

In this chapter we shall be considering situations in which the smooth course of behaviour is interrupted in various ways. Firstly, an animal may be trying to achieve some result but its attempts are thwarted. A hungry rat can see food at the end of a runway, but finds its path blocked by a glass plate. A bird trapped in a room makes futile attempts to escape through the window panes. Secondly, we have situations where two mutually incompatible tendencies are aroused simultaneously. Suppose the hungry rat in the runway can see food and its path is not blocked, but it has previously received an electric shock at the place where the food is. The food stimulates it to approach, but memory of the shock keeps it back.

We might label the first examples as 'thwarting' where only a single tendency is aroused, and the latter 'conflict' where two tendencies oppose one another. In practice there is a good deal of overlap, and it has been suggested that 'pure' thwarting does not exist since frustration may lead to avoidance of the frustrating situation and thus a kind of approach/avoidance conflict may be set up. At the moment we have insufficient evidence to decide this, but in any case thwarting and conflict can be considered together because both have similar behavioural results and both tend to produce the same physiological changes in the body.

Conflict is an important topic because, contrary to popular belief, it is not only man who is exposed to frustration and torn by opposing aims. Things do not always run smoothly in the animal world and, as we shall see, natural selection has produced a number of ways for dealing with conflict situations. In general, conflicts in nature are of rather short duration, but it is easy to produce a chronic conflict situation in a laboratory animal. Such situations have been extensively studied by psychologists because they may throw light on human neuroses, some of which may originate from just such a chronic conflict.

Here we shall concentrate on 'natural' conflicts in their biological setting, dealing mainly with the work of ethologists on vertebrates, but in conclusion we shall try briefly to relate this to some of the work on animal neuroses.

'Stress'

At this point it is necessary to consider the bodily changes which are likely to occur in conflict situations. We can lump them under the term 'stress', because the body's response to a wide range of 'stressors' (thwarted escape, overcrowding, extreme cold and burns, for example) is very similar. Barnett[15] describes the physiological changes involved and points out that most of them are attempts to restore the delicate balance of the body's metabolism when it has been upset.

In moderate stress we can detect increased activity in the autonomic nervous system (see Romer[216] for details of its structure) which supplies the viscera and smooth muscle. It also supplies the adrenal glands, endocrine organs close to the kidneys which have a double structure, an internal medulla (supplied by autonomic nerve fibres) and an external cortex. Stimulation of the adrenal medulla via its autonomic nerve supply causes it to release the hormone *adrenalin* into the bloodstream. This causes changes in numerous parts of the body; the sweat glands of the skin begin to secrete, hair becomes erected, the heart beats faster, breathing becomes more rapid and deeper, and blood gets diverted to the muscles from the alimentary canal. We have already mentioned these changes in the last chapter, because they also accompany the strong arousal of tendencies such as attack, escape and sex; they prepare the body for violent action of any type required. In brief conflicts there will perhaps be a rapid flush of adrenalin through the animal which then subsides, but if the stressful situation persists, then a further reaction begins. This involves the adrenal cortex which is stimulated to release its hormones, not directly by nerves as is the medulla, but by another hormone, *adreno-cortico-trophic-hormone* (ACTH) produced by the pituitary gland. Here, as with the release of adrenalin, the nervous system initiates the response. Stress activates cells in the hypothalamus which itself then stimulates the pituitary to release ACTH. It is not yet fully clear how the adrenal cortex hormones help an animal to adapt to stress. Some of them are concerned with glucose metabolism and may serve to mobilize the body's long-term food reserves. There is now some behavioural evidence that ACTH itself helps to reduce 'anxiety'. Whatever their action, the release of adrenocortical hormones is most dramatic. The cortical cells become drained of their contents and, if stress persists, the adrenals enlarge, sometimes by 25%, and all this growth is found to be in the cortex.

Animals under chronic stress become really ill and may die. Barnett[15] has shown how a wild rat, unable to escape from the territory of a dominant, resident male, may die after a few hours of intermittent attacks, even though it has no wounds. Persistent stress due to overcrowding has been suggested as one reason why natural populations of voles may show a rapid decline after a period of unusually high numbers.[50,54]

The diverse physiological changes we have been discussing can be regarded as successive stages on a scale of 'arousal'. In Chapter 3 we mentioned how any incoming stimulus leads to the activating of the brain's higher centres via the reticular formation; at this stage an animal will begin to 'pay attention'. If the stimuli are strong and persistent the hypothalamus initiates activity in the autonomic nervous system; adrenalin begins to circulate and the animal becomes highly aroused and perhaps shows 'emotion'. The final stages on this scale are those just described, with intense and persistent arousal leading to the release of adrenocortical hormones and the animal showing signs of severe stress (see Hokanson[117C] for further details).

We shall find that a knowledge of the underlying physiological changes helps in the interpretation of some behaviour which appears in conflict situations.

Territorial conflicts

Perhaps the best way to approach the study of conflict as it occurs naturally is by way of territorial behaviour which, as we shall see, almost inevitably leads to conflict where one territory meets another. The best definition of territory is 'any defended area'. Commonly, at the beginning of their breeding season, males establish and defend an area against other males of the same species. Such behaviour is widespread among the vertebrates. Many fish and lizards are territorial, so are the great majority of birds and many mammals; amphibians are the only vertebrates which show little territorial behaviour. Territory crops up sporadically among invertebrates in such insects as crickets and dragonflies and in the fiddler-crabs, but it is relatively rare.

We are still far from a complete understanding of the ultimate functions of territory. Something whose establishment and defence takes up so much of an animal's time must have a powerful selective value. A host of possible functions have been discussed, the majority of which suggest that this spacing out of the population is related to the available supply of food. The exact nature of this relationship is still a matter of dispute and the various arguments can be followed in Lack,[143] Wynne-Edwards[287] and also in Volume 98 of the ornithological journal, Ibis, which contains a series of articles on the rôle of territory in various bird species. There is some further discussion of this problem in relation to social organization in Chapter 10.

Certainly the relationship between territory and food cannot always be the same, because the form which territories can take varies so greatly. Some birds of prey defend a large area a mile or more square, from which they get all their food. The herring-gull's territory is a couple of square yards around its nest in the gull colony. In the bird of prey, food and territory are directly related, but the gull gets no food from its territory. However, its territorial behaviour will limit the size of a gull colony and thus control the number of herring-gulls feeding on the nearby sea-coast.

Apart from food supply, there are other selective advantages to holding a territory which in some species may have played an important part during evolution. Among them are freedom from disturbance during pair forma-tion and the avoidance of predation and disease.

Here we must concentrate on the behaviour associated with holding a territory. The first essential for a territorial animal is that it must be aggressive towards others of its kind. As we discussed in the previous chapter, aggression is seen in other situations, as when animals fight over food or establish a 'peck order' within a group, but it is most conspicuous at the start of the breeding season when territories are being established.

Aggression alone is not sufficient. If animals were simply aggressive they would spend far too much time and energy in fighting and damage them-selves. Further they might behave too aggressively towards their mate or their offspring as well as to a rival. Snow[248] records the case of one male blackbird who was extremely aggressive early in the breeding season and won a large territory. He attracted a female, but instead of the rapid waning of aggression between members of a pair which normally occurs, this male continued to attack his mate for over a month. His attacks were so persistent that she failed for several days in her attempts to start a nest.

Clearly this is aggression carried to a maladaptive level and we can observe an even more extreme case in one breed of the domestic chicken. Cock-fighting is still a popular sport in many parts of the world where deliberate selection for fighting abilities has been practised for centuries. Game-fowl have extremely powerful leg muscles, long spurs and a greatly reduced comb—a vulnerable spot for damage in normal breeds. They have the temperament to match their physique and are fantastically aggressive. A normal rooster on his home territory will attack and drive off any new-comer who is placed there—the latter normally beats a retreat with no delay. However, a game-cock placed on the territory of a dominant cockerel of another breed will instantly attack the territory holder and drive him off. The inevitable result of this abnormal aggressiveness is that game-fowl are difficult to breed, because females are aggressive too and much time is wasted in fighting.

Game-fowl would have little hope of surviving in the wild, where natural selection has prevented aggression from reaching too high a level and has favoured the evolution of a compensating tendency—that of escape. The majority of animals show escape responses to predators, but here we are concerned with escape evoked by a territorial rival.

After a few days of strife in early spring a population of male birds will begin to settle down, each with a territory in which he is dominant. Males which appear later and try to establish themselves will have to fight hard

Fig. 5.1 A territory map of a population of willow-warblers in birch woodland. The size of territories varies from under 1,000 square yards to more than 5,000 square yards. Shaded areas were unoccupied. (After May,[182] 1949, *Ibis*, **91**, 24.)

and may win only a small area. Territories have been likened to elastic discs —the more they are compressed, the more they resist further compression. Each male is most aggressive near the centre of his territory. As he moves away from the centre his attacks upon a trespassing neighbour are less violent, and there comes a point at which he is equally likely to attack or escape when his neighbour approaches. This point we can call the boundary of his territory and careful observations on a population of males enable one to construct a map with boundaries drawn through the points at which neighbouring birds show this balance between attack and escape. Fig. 5.1

shows such a map for a population of willow-warblers; similar maps have
been made for a group of sticklebacks in a large aquarium.

In summary, the establishment of a territory involves both attack and
escape behaviour, and if an animal meets its neighbour at the boundary
between their territories, both attack and escape tendencies are probably
aroused and form the basis of a conflict situation.

In such a situation an animal may do any one of a number of things. (The
term 'agonistic behaviour' is often used to cover all the different types of
response seen in fighting and territorial behaviour generally.) We may
observe actual attack or escape or their intention movements; the animal

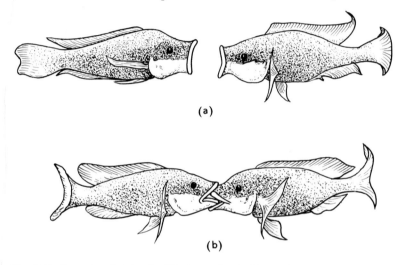

Fig. 5.2 Boundary fighting in *Tilapia natalensis*. The rivals approach with open
mouths displayed (a), and then pull with jaws interlocked (b). (From Baerends
and Baerends-van Roon,[11] 1950, *Behaviour Suppl.*, **1**, 1.)

may alternate between the two or show some kind of compromise movement
or posture.

Some types of agonistic behaviour have become modified by evolution
into displays which probably serve both to intimidate a rival and to reduce
actual fighting to a minimum. One of these displays—'threat'—we shall
discuss in some detail, but first we may consider fighting itself.

'Pure' attack and escape

During fighting we can observe attack and escape behaviour, each operating to the virtual exclusion of the other. Prolonged fighting is extremely
rare under natural conditions, the weaker of two combatants usually

breaks away and escapes before any serious damage is done. Clearly
selection favours individuals who quickly recognize defeat. 'He who fights
and runs away, lives to fight another day.' Sometimes fighting behaviour
has evolved into an almost formalized contest in which two rivals can assess
each other's strength without damage. Lorenz[160] describes the mouth-
fighting of various cichlid fishes. In *Tilapia natalensis*, rivals approach each
other displaying wide open mouths, seize their opponent's jaw and then
pull against each other (Fig. 5.2). In the related *T. mossambiqua* the rivals
oppose their opened mouths and push. In each case the weaker fish soon
breaks away and accepts defeat.

Other cichlid fish show 'pendulum' fighting at the boundary of their
territories, with attack and escape rapidly alternating. This is also familiar
in birds; one male flies unhesitatingly at an intruder who flees, but the
momentum of the chase carries both birds over the boundary between
their territories and promptly their behaviour changes with pursuer
becoming pursued. The rapidity with which attack can succeed escape in
pendulum fights demonstrates how closely both tendencies are dependent
on the external situation.

Quite commonly a victorious animal whose rival has been driven off, will
remain highly aggressive for a time. In such a state he may show the redirected aggression, which was discussed in the previous chapter and
attack another neighbour or even his mate, who would not normally
become involved.

THREAT DISPLAYS

'Compromise' behaviour of various kinds may be seen when the tendencies to attack and to escape, or, more generally, to approach and to
avoid, are both aroused simultaneously and to about equal extents. An
animal may remain rather still in one place, perhaps making abbreviated
intention movements to advance or retreat. Sometimes, instead of facing
one way or the other it will turn its body sideways, though the head may
still move from side to side. Behaviour of this type can be seen if one walks
slowly towards a bird when it is feeding young in the nest. A similar
conflict is shown by an octopus which has been given electric shocks on a
number of successive trials when it attacked crabs at one end of its tank. It
approaches the next crab in a hesitant, sidling fashion; half its tentacles
remain attached to the shelter in which it lives, the other tentacles
reach out towards the crab, but they approach it laterally rather than
directly. The octopus expresses most vividly a compromise between
advance and retreat in the actual way it holds its body.

Many territorial animals have special postures which are most commonly seen during boundary disputes. For example, two neighbouring black-headed gulls may approach one another but do not fight. Each adopts the rather 'forced'-looking posture shown in Plate 1, and may stand facing its rival or turn sideways and walk parallel to him. In a similar situation two male cichlid fish, *Cichlasoma meeki*, slow down as they approach one another and remain facing with their gill covers fully raised (Fig. 5.3a). Two hostile male rats turn sideways when they meet and twist around each other with arched backs and partially raised fur (Fig. 5.3b).

Ethologists call behaviour of this type 'threat', and consider it to be a direct result of a conflict between attack and escape tendencies when neither can find separate expression.

(a)   (b)

Fig. 5.3 Some threat postures. (a) Frontal threat display of *Cichlasoma meeki* (right) compared with the normal front view (left). (From Baerends and Baerends-van Roon,[11] 1950, *Behaviour Suppl.,* **1**, 1.) (b) Lateral threat display between male rats. (From Barnett,[14] 1963, *A Study in Behaviour.* Methuen, London.)

There are at least three lines of evidence for this conclusion.

1. Threat is most commonly observed at the boundary between territories where, as we have seen, there is good reason to believe both tendencies are aroused simultaneously. This might be called evidence from situation or context, and we can couple with this the observation that threat behaviour is most commonly preceded and followed by 'pure' attack or escape. This again, is what would be expected if these two tendencies were involved in its performance. Presumably a shift in the threshold of one or the other can lead to the often rapid transitions in behaviour which are observed.

2. Linked with this conclusion, there is some evidence on the causation of threat which comes from independent manipulation of the levels of

attack and escape tendencies. Blurton-Jones[32B] had a group of completely tame Canada geese which ignored him when he was dressed in familiar old clothes. If he wore a white coat the geese attacked him uninhibitedly, whilst if he appeared carrying a broom (used to drive the geese into their house for the night) they would flee. The familiar threat postures of the goose (lowered head on outstretched neck, hissing etc.) only appeared when Blurton-Jones combined wearing a white coat with carrying a broom!

3. Finally there is also evidence gained from a close examination of the form of threat postures which can sometimes be analysed into elements belonging both to attack and to escape behaviour, and for this reason such postures are called 'ambivalent'.

This last statement requires some amplification and, as an example, we can examine the lesser black-backed gull's threat posture (Plate 1a) more closely. The bird moves towards its rival with the neck stretched upward and slightly forward and the head and bill turned down. The wrist joints of the wings are lifted well clear of the body and the plumage is slightly raised. From a thorough knowledge of the gull's whole repertoire of behaviour it is possible to identify several elements of attack behaviour here. Gulls normally launch an attack by beating with the wings and attempting to peck down at their opponent. The raising of the head and down-pointing bill look like the first stages of actual attack. When a gull takes off to fly, the wrists are raised only at the last moment, so the raised wrist joints in the threat posture are probably attack also. But the gull does not attack, something holds it back and elements of escape can also be seen, particularly as two rivals come very close. Now the head moves increasingly back, the bill becomes lifted and the plumage sleeker, further the bird may turn sideways on to its opponent and move parallel to, not towards it. This turning aside looks like an element of escape and gulls draw back their heads and sleek the plumage preparatory to taking off, so these may be escape elements also. In fact this 'upright' threat posture, as it is called, can take on a range of forms depending on how many attack and escape elements are shown. This range is assumed to reflect slight changes in the levels of the underlying attack and escape tendencies. Fig. 5.4 shows three gradations of the 'upright' in the herring-gull which can be called 'aggressive', 'intimidated', and 'anxiety upright' respectively, as they show increasing elements of escape.

All this analysis of the threat behaviour of gulls is derived from Tinbergen and his students, who have made gulls one of the best-known groups of birds. Tinbergen[263] provides an excellent review of this work.

Not all threat postures can be interpreted as a mixture of attack and escape elements. Some of them include elements which seem to have their

(b)

(a)

Plate 1 (a) The 'upright' threat display of the lesser black-backed gull. (b) The 'hunched' appeasement posture of the same species (see p. 111). It represents an almost perfect antithesis to the threat posture. The head is held low on a shortened neck and the bill points upwards, whilst the wings are pressed close into the flanks so that the wrist joints—so conspicuous in threat—are completely hidden. (Photos by N. Tinbergen).

origin in the physiological side effects of conflict. The secretion of adrenalin and arousal of the autonomic nervous system will, as we have seen, lead to deeper and more rapid breathing. Fish will thus have to raise their gill covers to a greater extent in a conflict, and air-breathing vertebrates fill their lungs more. Morris[194] suggests that here is the origin of threat displays which emphasize the gill covers, as in *Cichlasoma*. Other animals, including some amphibians, reptiles and birds, inflate air sacs when they display. For example, the tropical frigate-birds have air sacs on their throats. These are normally quite inconspicuous but in display they become enormous and are bright scarlet in colour. In birds and mammals erection of the hair or feathers is another important result of autonomic activity in a conflict. Very many threat displays have this for a basis; birds have evolved conspicuous crests or ruffs which are raised in threat, mammals erect their hair, as in the familiar threat postures of cats and dogs. Morris provides many other examples of the way in which selection has produced striking displays based upon the body's physiological response to conflict.

Many boundary disputes go no further than mutual threatening followed by mutual withdrawal. Threat forms an admirable substitute for actual fighting and one writer has neatly summed up its rôle in animal disputes. It is not, 'speak softly, but carry a big stick', but rather. 'shout your head off, but run away if the other fellow isn't impressed'!

There is one complication to threat behaviour which is of considerable theoretical importance. We have earlier described the different forms of a threat posture which may correspond to changing degrees of escape tendency (Fig. 5.4), but some animals have several distinct threat postures which do not blend into one another. If we accept the general conclusion that threat occurs when attack and escape balance, how can this interpretation be applied when there are several types of threat?

One possibility is that the different postures represent different levels of the attack and escape tendencies. For threat to occur at all both must be roughly equal, but if both are low posture A appears, if both are high, posture B. On p. 102 we discussed the observation that 'pure' attack and escape were the most common sequel to threat behaviour. This fact might assist in estimating the causation of different threat postures by watching the outcome of an animal adopting postures A and B. Assuming—and it is only an assumption—a particular type of interaction between attack and escape tendencies, one might expect that if threat posture B occurs when both tendencies are high, then it would be more commonly followed by attack or escape than would posture A. Because both tendencies are low in A, other types of behaviour—feeding or preening, for example—might be able to occur because their control mechanisms would have less 'resistance' to overcome from attack and escape tendencies.

Tinbergen,[263] whose group have used this method, recognizes that it has limitations. Under field conditions it is very difficult to be sure that a bird's behaviour following threat is not affected by the responses of others. Further, as we shall discuss later, there are other types of behaviour (displacement activities) which result from a conflict, and these might appear

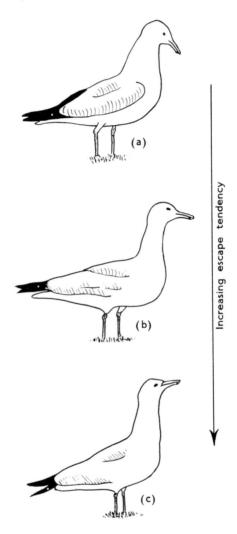

Fig. 5.4   (a) 'Aggressive', (b) 'intimidated', and (c) 'anxiety' upright threat display of the herring-gull. (From Tinbergen,[263] 1959, *Behaviour*, **15**, 1.)

following threat and upset the kind of analysis outlined above. Moynihan[196C] has used this method with the black-headed gull which has four distinct threat postures called, 'upright' (see Plate I) 'oblique', 'forward' and 'choking'. He interprets their causation in the terms outlined above, with 'choking' representing the highest levels of conflicting attack and escape. Stout[251A, 251B] and his collaborators have worked with the glaucous-winged gull which has a very similar repertoire of threat postures. Like Moynihan they also rank the different postures by observing the proportion which lead to attack. They also looked at the number of attacks made by a territory owner when an intruder adopted each type of threat posture. They suggest that only the tendency to attack is varying, although presumably they would accept that some conflicting tendency inhibits its expression during posturing.

Recent work on gulls and other birds indicates that tendencies other than attack and escape may be involved in the performance of postures which hitherto have been lumped together as 'threat'. 'Choking' in gulls, which has just been mentioned, is a case in point. It is performed most commonly by birds on their territories which are very reluctant to move, and its form resembles nest building in some respects. It seems reasonable to conclude that a rather non-specific tendency 'to stay put' is involved here. Blurton-Jones[32C] came to a similar conclusion from his detailed analysis of displays in the great tit, where, although all of the threat postures depended on the presence of a stimulus which elicited attack tendencies, other tendencies determined which posture appeared. 'Head-up', for example, (Fig. 5.5a) certainly involves an escape tendency interacting with attack, but 'horizontal' (Fig. 5.5b) seems to depend, like 'choking' in the gulls, upon a strong tendency to stay put.

In terms of the behaviour they produce, there can be no doubt that the attack and escape tendencies are mutually inhibiting. It is perfectly reasonable to suggest that the neural mechanisms behind the overt behaviour are also mutually inhibitory. As we discussed in Chapter 1, it is by inhibition that sudden but smooth transitions from one type of behaviour to another are achieved. The behavioural evidence would indicate that attack and escape tendencies can remain in a fairly stable balance and still allow the simultaneous expression of behaviour elements belonging to both. As yet we do not know how this is achieved in neurophysiological terms. In particular we have no idea how a balance between the two struck at different levels could produce different types of threat posture. Leyhausen[152] has proposed on behavioural grounds that there is a continuous range of threat postures in the cat which represent different relative strengths of attack and escape tendencies. The facial expressions associated with these postures are illustrated in Fig. 5.6. Brown and Hunsperger[39] have studied attack, escape and threat in the cat by implanting stimulating electrodes in the hypothalamus and elsewhere. They do not consider that attack and escape are distinct systems because the areas of the brain from which they can be elicited overlap with each other and with areas which produce threat. They found it impossible ever to get 'pure' attack by electrical stimulation, it was always preceded by threat, and they could elicit threat from single points of stimulation at widely different loci in the hypothalamus and amygdala.

As we discussed in the previous chapter, there are problems in interpreting brain stimulation experiments. Brown and Hunsperger's results cannot be said to exclude the possibility that there are distinct attack and escape systems which interact, but their suggestion that threat displays

Fig. 5.5 Variants of (a) the 'head-up' and (b) the 'horizontal' displays of the great tit, *Parus major*. (From Blurton-Jones,[32C] 1968. *Anim. Behav. Monogr.*, **1**, 75.)

may have a more unitary basis is an important one. We can retain the conflict hypothesis at a behavioural level because of its usefulness in descriptive studies, but we must await further research for its physiological interpretation.

Threat as a signal

An experienced ethologist can usually predict what an individual animal will do next from its behaviour at a given moment. The analysis of the upright threat posture of gulls is based on a full knowledge of their behaviour which enables the intention movements of attack and escape to be identified. If we can interpret intention movements, it seems highly probable that other members of the same species can do so. We have seen

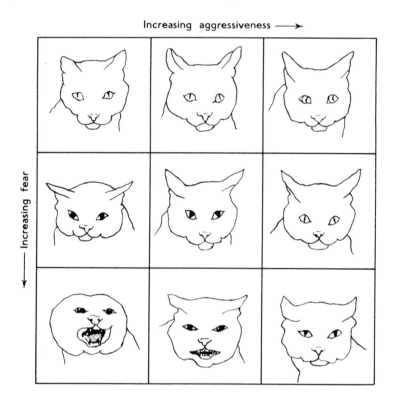

Fig. 5.6 Changes in the facial expressions of the cat with their suggested motivational basis. (From Leyhausen,[152] 1956, *Z. Tierpsychol., Beiheft,* **2**, 1.)

that many threat displays are made more conspicuous by specially evolved releaser structures which emphasize the movement or posture. Threat postures are, in fact, a primitive form of language by which simple information on the motivation of the displaying animal is passed on to others. Such displays are often called 'social signals' and Marler[179] discusses their function and evolution in more detail.

In practice it is rather difficult to measure what information is conveyed by a threatening animal to another at which it is displaying. Stout's[251A, 251B] study of the glaucous-winged gull included work with dummies mounted in various threat postures and placed in a pair's territory. This situation is rather unnatural, in that a gull instead of being approached by an intruder from the edge of its territory, returned from an absence to find the intruder already in occupation, as it were. Stout found that postures which on other grounds (see p. 107) were deemed to be most aggressive, were attacked less often and with longer latencies than less aggressive ones. The presumption was that a posture conveyed its aggressive content to the territory owners who attacked most vigorously when they had least fear of retaliation. Stout and his collaborators found that the level of the head above the ground was all-important in conveying aggressiveness—the lower the head, the more aggressive the posture. Thus dummies in the 'upright' posture were attacked with least hesitation and those in the 'choking' posture with most reluctance. Interestingly enough, when a loudspeaker was used to emit calls from the dummies, the rhythmic call that accompanies the choking posture was most effective in increasing the latency of attack upon all dummies whatever their posture. No other call was so effective and clearly the choking call also functions a highly aggressive signal.

One of the most detailed analyses of the communicatory rôle of threat is that of Stokes[251] on the agonistic behaviour of blue tits. He observed them disputing over food in winter and recorded various elements in their threat postures, involving beak, crest, body feathers, wings, tail and orientation of the body, among others. Stokes does not choose to ascribe individual elements to attack or escape, he considers they may all reflect both tendencies. Particular combinations of elements were much commoner than others and sometimes provided a very reliable prediction of a bird's subsequent behaviour. Thus blue tits which faced their rivals with raised crests subsequently showed escape in 94% of cases and they never attacked. On the other hand, attack was much harder to predict; no combination of elements gave a probability greater than 48%.

Stokes measured the responses of one bird to a rival which adopted a particular posture, eliminating those cases when the latter actually attacked. If a bird, A, adopted an 'attacking' combination of elements this significantly increased the frequency with which its rival, B, would escape— although as we have just seen B may have escaped unnecessarily. If A adopted an 'escape' combination, B's frequency of attack did not rise but it tended to ignore A. It probably recognized that A's posture meant that the latter did not constitute a real threat to its position at the feeding dish. We badly need more quantitative information of this type to augment the wealth of descriptive data on threat behaviour which we already have for a number of species.

Appeasement displays

It is most convenient to mention appeasement displays here because they are often related to threat and are a further mechanism for reducing fighting. Functionally speaking, their rôle seems to be that of reducing or inhibiting attack in conditions when escape is disadvantageous. We have mentioned that an aggressively aroused male may redirect his attack upon his mate, but it may be disadvantageous for her to flee out of his territory; it is better to stay put. Young birds may also become the victims of redirected attacks and use appeasement rather than escape. Some social animals such as wolves, baboons and domestic chickens form a stable hierarchy or 'peck order' among the members of a group (see Chapter 10, p. 248). Once this is fixed, fights are rare and dominant animals have only to threaten or even move towards a subordinate and the latter defers, often with an appeasement gesture. Again, it is usually disadvantageous for a subordinate animal to escape by leaving the group.

At present our evidence on the functioning of appeasement rests entirely on behavioural descriptions, but it is none the less quite convincing. Appeasement seems to operate in two different ways. Firstly, attack can be inhibited by arousing a conflicting tendency in the attacker. A subordinate baboon of either sex turns away from the aggressor and crouches in the sexual presentation posture. Sometimes it may be mounted briefly, but often it is simply allowed to move away by the dominant baboon which 'accepts' this as a gesture of submission. Among birds it is quite common to find infantile behaviour used for appeasement. Food-begging behaviour is a common preliminary to mating in gulls and some other birds, and it appears to reduce a male's aggressiveness, perhaps by arousing his parental responses.

The second type of appeasement relies on being as different from threat as possible. Threat postures emphasize various releasers, but appeasement postures hide them. In the black-headed gull all of the threat postures emphasize the dark face mask. When a female alights on the territory of her mate they may threaten each other briefly, but after a second or two both lift their heads and jerk them away from each other—hiding the face mask. The kittiwake threatens with the yellow bill, often opened to display the vivid red mouth lining. The appeasement posture turns the head away from the opponent and down, hiding the bill. Plate 1b illustrates the appeasement posture of the lesser black-backed gull, which is of this type and an almost exact antithesis of the threat posture shown in Plate 1a. In fact, this example bridges our two types of appeasement, because this anti-threat posture is also that adopted by young birds when food begging, and is used by adults in courtship.

There are examples of anti-threat appeasement postures in mammals

also. Wolves threaten by snarling with bared fangs and erect ears. A
defeated individual, or one who wishes to appease a more dominant wolf
in the hierarchy, lays its ears back and turns away its head so as to
situation in wolves which threaten by snarling with bared fangs and erect
ears. A defeated individual, or one who wishes to appease a more dominant
wolf in the hierarchy, lays its ears back and turns away its head so as to
present not its fangs, but the nape of its neck; domestic dogs often do
precisely the same when being punished. *dogs rolling over on to back etc.*

Appeasement displays, just as threat, act as signals between animals. We
are left with the problem of why appeasement should be effective. The
appeaser is vulnerable and yet the dominant animal rarely attacks further.
It is irrelevant to suggest that animals feel mercy—in fact taken out of their
natural environment and crowded together they will often kill each other.
We assume that selection favours the inhibition of aggression when
appeasement is offered. Animals which do not are likely to be hyperaggres-
sive and as maladaptive as the game-cock. Baboons, for example, must live
in a group to survive, because stragglers are usually killed by leopards or
lions. Selection favours any behaviour, such as the appeasement system,
which helps to keep the group together. Chance[49] has pointed out that
appeasement displays may serve to reduce stress in the submissive per-
former because, by turning away from the dominant animal, it minimizes
the 'frightening stimuli' which impinge upon it. In this way the escape
tendency may be reduced sufficiently so that the submissive animal is able
to stay with the group.

DISPLACEMENT ACTIVITIES

We must return to conflict situations and discuss some other types of
behaviour which are observed. The name 'displacement activities' was
given by Tinbergen[259] to a diverse range of behaviour patterns whose most
striking common characteristic is their apparent irrelevance to the situation
in which they appear.

For example, in the middle of a bout of threatening interspersed with
fighting, two cockerels will turn aside briefly and peck at the ground, sometimes picking up stones or grains which they allow to fall again. A male
stickleback which will be courting an unreceptive female will suddenly
swim to his nest and perform the characteristic parental 'fanning' movement which ventilates the nest with fresh water, even though there are no
eggs present. A tern which is incubating on its nest makes a few brief
preening movements just before it takes off at the approach of an intruder.
A thirsty dove which is prevented from getting to its water bowl by a
sheet of glass, pecks at the ground nearby.

angry employer hits desk instead of secretary.

In all these examples there is good reason to suppose that the animal is either thwarted or in a conflict between two opposing tendencies. The appearance of a brief burst of, say, feeding in the midst of bouts of threat and fighting is surprising, because it seems irrelevant to the tendencies aroused—presumably attack and escape. Normally animals are consistent and one type of behaviour continues for some time without interruption. Further the preening performed by the tern just before it leaves the nest often appears 'forced' and incomplete. So does the feeding of the fighting cockerels, which, as we mentioned, may pick up grains but do not swallow them.

The obvious subjective conclusion is that the animal is under 'tension' when performing these acts. Here is one case where subjective experience, if used with caution, may help in understanding animal behaviour. We all know that when we are ill at ease we are apt to do completely irrelevant things without much conscious thought. People rarely sit peacefully in a dentist's waiting-room or outside an examination hall. They fidget, making minor adjustments to necktie or hair, light cigarettes and talk unnecessarily fast and eagerly. It would be perfectly feasible to obtain a measure of the 'stickiness' of a cocktail party, from the amount of potato crisps and nuts eaten during its early stages. Nobody is hungry, often quite the reverse, but eating in some way relieves the tension of being with strangers.

Associated with these human examples there are often signs of physio-logical arousal with raised autonomic activity. This produces a quickened pulse, sweating and the familiar effects of psychological tension on bowels and bladder. Sometimes we can detect similar changes in animals when they perform apparently irrelevant movements.

The term 'displacement activities' refers to a particular theory of their origin which we owe to Tinbergen and Kortlandt (see Tinbergen[259]). This suggested that in the normal control of behaviour a particular quantity of 'nervous energy' is released for the performance of a behaviour pattern, A. The release of this energy through the normal A outlet can be interrupted in two ways. The correct external stimuli for A are absent or some other pattern, B, is aroused whose performance is incompatible with that of A. If this interruption occurs, the energy which pertains to A must find an alternative outlet, and it is 'displaced' through to channel C; behaviour pattern C then appears as a displacement activity. Tinbergen's theory grew out of a 'psycho-hydraulic' type of motivation model, such as we have previously discussed in Chapter 4 (Zeigler[292] provides a clear review of the whole displacement activity concept which sets this theory in per-spective). It has provided a useful stimulus to research, but has proved inadequate on a number of counts.

The theory set up something of a distinction between a pattern—say preening—performed normally and the same pattern performed as a displacement activity. In the former case preening would be using up its own supply of nervous energy, whereas in the latter the preening control mechanism would be receiving the displaced overflow from another pattern. This implies that preening may occur as a displacement activity without there being any previous activation of its control mechanism, either from within or without; i.e. in the absence of any tendency to preen or external stimulus to preening.

Careful observations have not supported this contention and indicate that the control of a behaviour pattern, and the influence which external stimuli have upon it, is the same no matter in what circumstances it is performed. For example Räber[208] found that turkey cocks showed bouts of displacement feeding or drinking during fights; which pattern appeared depended on whether food or water was available to the birds. Van Iersel and Bol[128] studied nesting terns which were in a conflict between the tendency to stay on their nest and the tendency to escape. They found that the amount of displacement preening increased when the terns had got wet feathers—a normal stimulus to preening. Sevenster[237] mimicked the effect of eggs developing in the nests of sticklebacks by passing through water containing dissolved carbon dioxide. This increased the amount of displacement fanning a male performed during bouts of courtship.

Sometimes the thwarting or conflict situation which leads to the appearance of displacement activities may also enhance the relevant stimuli for them. Deaux and Kakolewski[61A] have shown that a brief period of stress caused by handling rats, led to abrupt changes in the salt concentrations in the body fluids—one of the normal signals of water deficit—and consequently caused them to show a brief period of drinking. This might well be labelled as 'displacement drinking' by an ethological observer. The physiological arousal caused by a conflict situation will cause sweating and erect of hair or feathers, any of which is likely to produce skin irritation. It is probably not chance that preening and grooming appear so commonly as displacement activities. Andrew[8] pointed out that the feather postures of birds in a sexual conflict situation resembled those of birds which he had placed in a very hot room and which were trying to cool themselves.

Any quantitative measurements of displacement activities require strict controls. One must not label some pattern a displacement activity without good evidence. This is particularly true when something like preening is involved, because birds preen a great deal and in a variety of situations, conflict and otherwise.

Some experiments by Rowell[224] show what can be done. He worked with chaffinches in aviaries and used two methods to produce an approach/avoidance conflict. In one, a stuffed owl was placed outside the aviary which evoked the chaffinches' mobbing behaviour, described in Chapter 1. The aviaries had a row of perches which allowed the birds to settle at varying distances from the owl. The second situation used hungry birds which had previously been trained to use a food dish at one side of the aviary. During these tests, when they reached the dish, a bright light was flashed on inside it which provoked escape. In both cases the birds tended to make short flights between perches, resting briefly on each and gradually getting closer to owl or food dish, but subsequently retreating. After a pause on a more distant perch they gradually advanced again, and so on. With the owl we might consider the bird was in a conflict between attack and escape, whilst with the food dish, hunger and escape were conflicting.

A pause on any perch represents a brief period of stability when the bird's tendency to go forward and that to go back balance. Rowell measured the length of pauses and found that, in both conflict situations, the birds paused more often and for longer at intermediate perches, neither too near nor too far. This part of the aviary is analogous to the boundary of a territory in many ways—the region of maximum stability. It is here that we might expect displacement activities to be most frequent.

Rowell did find that chaffinches showed most preening and bill-wiping during their pauses on the intermediate perches. But one would get the same result if preening simply occurred at a fairly constant rate, wherever a bird made a pause and was doing nothing else—the longer the pause, the more preening.

In control observations of birds under no conflict Rowell found this was indeed the case, but from these controls he could calculate how much preening would be expected for pauses of a given length. The amount of preening per length of pause was significantly greater when the birds were in a conflict, which justifies labelling this extra amount 'displacement preening'. Wetting the bird's feathers tended to increase preening in both conflict and non-conflict situations, so Rowell's work agrees with that quoted earlier in showing that displacement activities are facilitated by external stimuli.

Van Iersel and Bol[128] recognize this in their 'disinhibition' hypothesis for the mechanism of displacement activities. This is represented diagrammatically in Fig. 5.7. 'Centre A' controls the performance of behaviour patterns A, 'Centre B' those of B. The two centres are mutually inhibitory, represented by the minus arrows, so that patterns A and B cannot be performed together. One or both of these centres has an inhibiting effect on a third 'Centre C'. Now suppose both A and B are aroused simultaneously, they will inhibit each other's activity and neither will be able to sustain their inhibition of Centre C which, if it is aroused, will initiate behaviour patterns C.

It is a simple matter to substitute the names of relevant tendencies from our examples for A, B and C in the diagram. We must remember that our justification for doing so is based only on behavioural observations and we cannot yet ascribe any physiological reality to the disinhibition system. In the case with terns, originally investigated by van Iersel and Bol, they suggest the incubation tendency is mutually inhibitory with escape. Preening, previously inhibited by incubation or escape or both, appears when they are in balance. In Rowell's experiments with chaffinches, either hunger or attack conflicts with escape and again it is preening which is disinhibited.

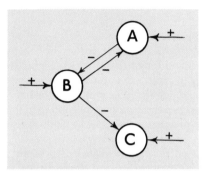

Fig. 5.7   A diagram of the 'disinhibition' hypothesis of van Iersel and Bol.   See explanation in text.

The disinhibition hypothesis can account only for displacement activities which occur as a result of a conflict between two incompatible tendencies. Tinbergen[259] originally suggested that the thwarting of a single tendency could also lead to displacement. More recently McFarland[164] has experimented with thwarting the drinking of thirsty doves, for example by placing a glass screen between them and water. 'Displacement pecking' occurred in this situation as well as in clear approach/avoidance conflict situations of the type used by Rowell.[224] Both Rowell and van Iersel and Bol[128] deny that pure thwarting exists. They assume that thwarting is punishing in itself and therefore leads to avoidance, but there is no real evidence that this is always so. McFarland[164A] discusses this problem more fully and suggests a rather different type of mechanism for displacement activities which attempts to draw together conflict and thwarting situations alike. In brief outline, McFarland's theory proposes that the brain is constantly comparing the results of the various behaviour commands that it produces, with an internal 'expectation' of their results. (This expectation may be built in to the nervous system during development or it may have to be learnt—it will certainly be subject to modification by experience.) We have already met the concept of a behavioural output being compared with an internal expectation when we discussed the development of bird song, p. 33. There the singing bird compared the sound it produced with an internal 'template' which effectively functions as an 'expectancy' of what the correct song should sound like. McFarland suggests, for example, that a dove approaching a situation in which it has previously fed has an analogous 'expectancy' of what the result should be. Normally the result and the expectancy match up—the bird pecks in a familiar food bowl and its crop fills with grain etc. There are two different ways in which this situation can go wrong. Firstly, the food bowl may be covered by a glass plate—a thwarting situation. Secondly the dove may have another simultaneous behaviour command to avoid the area near the food bowl because of previous unpleasant experience there. This second command will also produce a second expectancy and a conflict situation develops. In either case there will be a mismatch between the results of the animal's behaviour and the expectancy or expectancies. Then, according to the theory, a higher brain centre which determines the behaviour commands that are issued comes into play. So long as expectancy and results match, the behaviour command is stable and changes only when external or internal factors change; when expectancy ceases to be matched there is an increasing tendency for the higher centre to switch attention to another command and thus another type of behaviour—the displacement activity—appears.

McFarland's theory involves a type of disinhibition, because that is effectively what the switch of attention does, but it allows for disinhibition in thwarting situations as well as in conflict. It also allows for the accessibility of displacement activities to the normal external and internal factors that affect their performance in other situations. McFarland claims to be able to detect behaviourally the switch of attention that accompanies the performance of displacement activities. In a conflict or thwarting situation his doves either stand still, fixating their objective, e.g. the food bowl, or stand in a relaxed fashion, looking around. It is during the latter 'attentive' posture that they turn to preening or pecking at the ground.

Several questions remain concerning the outcome of conflict situations. For example, when there is mutual inhibition between attack and escape, what determines whether threat or, say, displacement preening occurs? Does it depend upon the absolute levels of attack and escape, or upon how closely they are balanced, or on some other factor? Some clues are suggested by Kruijt's[142A] observations in the jungle fowl, where several different behaviour patterns appear during fights between males. Kruijt found that the frequency with which the patterns were performed varied greatly depending on whether a bird was winning or losing. Thus to take the two extremes, pecking at the ground was more than four times as common in winners than losers, whilst the converse ratio was found for preening. It is reasonable to suggest that ground pecking results from a higher level of aggressiveness than preening, and that birds which preen are more motivated by escape. Should we regard these behaviour patterns as displacement activities resulting from different levels of conflict between attack and escape, or are they more direct reflections of single motivational states? Ground pecking might arise from redirected aggression in victorious birds and preening from the more stressed condition of a loser. The behavioural evidence is so far inadequate to provide an answer, nor can physiology help yet.

If a displacement activity is to appear it seems probable that the thwarting or the equilibrium between two conflicting tendencies must persist for long enough to allow its performance to 'break through'. For example, if one tendency is rising very rapidly with respect to another with which it conflicts, then there will be no time for a displacement activity to appear during the brief period in which equilibrium exists.

In this connection Rowell makes an important point about the rôle which disinhibition may play in the change over from one type of behaviour to another. He describes how chaffinches often show a cycle of behaviour, lasting about 20 minutes, in which periods of sleep or rest on a perch alternate with periods of feeding on the ground. Quite long bouts of preening occur, especially during 'waking up' when, presumably, the feeding tendency is slowly rising with respect to sleeping. Preening is probably the result of disinhibition here just as in the conflict situations we have been considering.

In other words, preening is a rather 'low-priority' activity. Plumage can be kept in order even if preening occurs only when a bird has no other strong tendencies. Consequently it will appear between periods when one or other of the stronger tendencies has control of behaviour. If a period is short, the preening may have the characteristics of a displacement activity, but if it is long, as in the example just quoted, we would usually call it 'normal preening'.

Displacement activities are probably just one example of a general feature of behavioural organization. There will be 'competition' for control of an animal's behaviour, and various factors, such as mutual inhibition between two tendencies, the external stimulus situation, and perhaps even the posture or movement which the animal is performing,[153] may facilitate another tendency taking over. If a tendency is increasing, it is more likely to succeed in taking over, whether there is conflict or not. McFarland[164] has shown that when doves are hungry they show more 'displacement feeding' behaviour in a thwarting situation and their behaviour will, of course, become increasingly 'food orientated' in all respects.

In conclusion it may be noted that, just as with ambivalent postures, natural selection has modified some displacement activities to serve as social signals. These are dealt with in the next section because all the best examples come from courtship behaviour.

COURTSHIP AS CONFLICT BEHAVIOUR

One of the most important contributions which ethology has made to the study of behaviour has been in the field of animal courtship. Courtship may be defined as 'specialized behaviour patterns which form the normal preliminaries to mating'. It occurs rather sporadically through the animal kingdom and is commonest in the arthropods and the vertebrates.

The function of courtship is usually to synchronize the reactions of male and female so that copulation can take place. Overall synchrony of reproductive behaviour is achieved by day length or temperature via the endocrine system, but a much more precise synchrony is necessary. This is because the close presence of a potential mate arouses, not only sexual tendencies, but also others which may be incompatible with sex. In some arthropods, such as spiders, this is seen at its most dramatic because a male is apt to be treated as prey by the larger female! His courtship is, in fact, a signalling system designed to establish his identity and, finally, to immobilize the female so that he can mate with her. The courtship of arthropods is, so far as we know, very different in origin and detailed function from that of vertebrates with which we are most concerned here. More detailed discussions of arthropod sexual behaviour can be found in Carthy[45-6], Evans[70B] and Manning.[177]

In vertebrates, because of the territorial system, a male's first response to a female may show elements of attack and escape in addition to those of sex. This holds true even when there is strong sexual dimorphism and it would seem easy to distinguish a female from a trespassing male. Sometimes the female is attacked, although often she does not flee but remains still and may take up an appeasement posture. We have already discussed an example of this type in the pairing behaviour of the highly aggressive cichlid fish *Etroplus* (see p. 83). In some birds the female becomes dominant after pair formation and the male's subsequent behaviour shows more overt signs of fear.

Attack and escape behaviour can also be seen at the time of courtship. The zig-zag courtship dance of the male stickleback is commonly interspersed with actual attacks on the female. Conversely, copulation in chaffinches is sometimes broken off when the female turns and attacks the male who flees. In animals which pair for any length of time, aggression tends to die down as they get to know each other, but it rarely disappears completely.

Lorenz[160] has stressed the importance of such aggression in the formation of the 'pair bond' in many vertebrates. This aggressiveness, although it may be aroused between the male and female, is largely redirected outwards towards neighbouring animals. Often male attacks male and female attacks female; their co-operation in the defence of the territory appears to strengthen the bond between them. Some birds have evolved elaborate mutual courtship displays to fulfil the same function. The elegant displays between male and female great-crested grebes, first described and analysed by Julian Huxley[125] more than fifty years ago, are of this type (see Fig. 5.8); their displaying reaches a peak in the early part of the breeding cycle when

Fig. 5.8 Male and female great crested grebes, *Podiceps cristatus*, performing one of their elaborate mutual displays. This one involves presentation of nest material. (After Huxley,[125] 1914. *Proc. zool. Soc. Lond.*, **1914 (2)**, 491.)

the pair are settling down together, although it persists to some extent all through the summer. Feeding of the female by the male is also a common feature of bird courtship, and obviously derives from the appeasement gestures, resembling the food-begging of young birds, which the female makes when she enters the male's territory (see p. 111).

The reduction of aggressiveness is certainly an important aspect of courtship in vertebrates. Coulson[54C] has shown that in the kittiwake—a cliff nesting gull—the nesting success of birds paired with their old mate is greater than that of birds of the same age which have changed mates. Significantly, the former lay their eggs sooner, and this may be because they settle down together more rapidly. Kittiwakes are highly aggressive and there is a high level of aggression within a pair at the beginning of the season (see p. 174).

We mentioned that courtship feeding obviously has its origins in the male's aggressiveness towards the trespassing female and her appeasement gesture. It is not surprising that we can identify other elements of attack and escape which have become incorporated into courtship behaviour. The courtship of male chaffinches and snow-buntings resembles escape behaviour because they do not face the female but run away from her with spread wings. The courtship of cockerels, on the other hand, is a 'waltzing' movement around the female and the same waltzing is seen during fights between two males. The zig-zag dance of the stickleback may have evolved from an alternation between attacks on the female and leading to the nest; sometimes, as mentioned above, the 'zig' part of the movement still finishes as an actual attack whilst the 'zag' may finish as leading.

From the descriptions given earlier there are reasonable grounds for suggesting that the courtship situation may involve a conflict between sex and attack or escape and courtship is often interspersed with short bouts of preening or other patterns which resemble displacement activities.

In some animals natural selection has apparently so modified movements which originally occurred as displacement activities that they now form a normal part of courtship. Morris[196] describes movements in the courtship of grass-finches which probably represent different stages in the modification of displacement bill-wiping (see further discussion in Chap. 7, p. 162 and Fig. 7.6). Some of the most beautiful examples are in the ducks where Lorenz[156] and his pupils have been able to trace the evolution of displacement preening and drinking movements into displays. Some species have evolved further than others in this respect, but in all cases releasers have developed which emphasize the effect of the movement. In the courtship preening movement, a drake's bill is drawn along brightly coloured feathers on the wing which is usually elevated (Fig. 5.9).

We do not fully understand how this or any other type of courtship display functions. Presumably selection has modified those patterns which most stimulate the female sexually or at least inhibit her from moving away. Courtship behaviour is a fascinating topic and its study is important for illustrating a number of behavioural principles. The reader is recommended to more complete accounts in books by Tinbergen[261] and Bastock.[18]

Fig. 5.9 Courtship preening in four species of duck; in every case the movement serves to emphasize bright markings on the wings. (From Lorenz,[156] 1941, *J. Orn., Lpz.*, **89**, 194.)

PROLONGED CONFLICTS AND 'EXPERIMENTAL NEUROSIS'

We mentioned at the start of this chapter that experimental psychologists have studied conflict in animals because of its possible relationship to human neurosis. The 'natural' conflicts we have been discussing rarely persist for long. When two tendencies conflict, one usually gains fairly rapid precedence over the other, and territorial or courtship conflicts are usually resolved one way or the other within a few minutes. By contrast some laboratory-staged conflicts are prolonged indefinitely and generally allow the animal no chance to escape from the situation.

A common technique has been to set up a stable approach/avoidance conflict of the type Rowell[224] used with chaffinches. Animals are given an electric shock when they approach food, for example. Masserman[181] trained cats to open a box for a food reward when a signal light flashed. Later, when the food box was opened the cats sometimes received a strong blast of air. Under these conditions the animals' behaviour often became severely disturbed. Some of the cats became hyperexcitable, others moped in corners for days on end. They nearly all showed signs of acute stress, with raised blood pressure, hair erection and gastric disorders.

A similar kind of disturbance can be produced by presenting animals with an unsolvable problem. Pavlov[204] described various 'experimental neuroses' in dogs he was using for his conditioned reflex experiments (see Chapter 8). These developed when a dog was rewarded for responding with, say, a leg movement to a circular patch of light, but punished it if responded to an elliptical one. Once this discrimination had been learnt, the elliptical patch was, over successive trials, gradually made more circular. There came a point at which the dog could no longer distinguish between the two. Faced with this situation, some dogs became violently agitated, others responded to every stimulus regardless of shape, others ceased to respond at all and fell asleep.

None of the behavioural disturbances just described has been analysed in ethological terms. Undoubtedly some of the behaviour patterns could be considered as displacement activities and are similarly produced, but the disturbance goes far beyond this. It was often months before Masserman's cats recovered, and then only if they were kept away from the laboratory.

We know that the stress which results from a prolonged conflict is often severe enough to produce physical damage. Animals develop gastric ulcers or tumours in the pituitary gland. In such cases it is impossible to regard the response as being in any way adaptive. The unfortunate animal finds itself in a situation to which there is no solution. It is not surprising that no mechanism exists to deal with chronic conflicts. It could provide little selective advantage in the natural world where escape from the conflict situation is almost always possible as a last resort.

6

Hormones and Behaviour

In Chapter 4 we discussed a number of ways in which an animal's behaviour is linked to the metabolic state of its body. Food and water deprivation result in behaviour which is designed to restore such deficits and we noted that together with behavioural responses there may also be hormonal ones; thirst leads to the release of antidiuretic hormone from the posterior pituitary. We must now consider the interaction between hormones and behaviour in more detail.* This topic deserves close attention because in many ways the endocrine organs, which produce the hormones, and the nervous system share the common functions of communication and co-ordination both within the animal and between it and the outside world. The hormones form a chemical message system which is probably as old as the nervous system. Indeed the one may have developed from the other in part. Throughout the animal kingdom we find neurosecretory cells in the nervous system. These are modified neurons which can pass special chemicals down their axons and into the bloodstream. Often these cells are clustered together to form glands, such as the corpus cardiacum of insects, which have close connections with both nervous system and bloodstream. The vertebrate pituitary gland develops from the fusion of neural and epithelial tissue and, as we described in Chapter 4, remains closely connected to the hypothalamus. The pituitary regulates by its various secretions all the other endocrine glands and is, in turn, regulated by the nervous system. This system of control enables environmental changes, picked up by the nervous system, to be matched by an appropriate hormonal response. The most familiar example is the control of breeding seasons in mammals or

---

* The reader is referred to a series of monumental reviews edited by Young[290]; those of greatest behavioural interest are in the second volume. Beach[24] also gives an important account of this field.

birds. Changing day length, perceived by the eyes, results in changed activity in the hypothalamus which stimulates the pituitary. This in turn secretes hormones which start the various growth changes in the body associated with the onset of breeding condition.

Throughout all their interactions, the functions of the two communication systems—endocrine and nervous—remain essentially complementary. The nervous system can only pass information by trains of nerve impulses. Its state can change very rapidly, but it is clearly less suited to transmit a steady unchanging message for a long period; it operates on a time scale from milliseconds to minutes. The endocrine system cannot respond so rapidly, but its cells can maintain a prolonged steady secretion into the bloodstream lasting for months if necessary. Moreover hormones can reach every cell in the body via the bloodstream, whereas the nervous system generally controls only the muscles.

Although the effects of adrenalin secretion, described in Chapter 5, are rapid they are exceptional, and usually when behaviour changes rapidly we can rule out hormonal influences; slow changes which persist for days or longer are more likely to have an endocrine basis. Early in its reproductive cycle a male stickleback goes through a period when it will both court females and attack rival males. Sevenster[237] has shown that, measured on a minute to minute time scale, when one tendency is high, the other is low. After a period of attack, for example, a male will not court a female for several minutes. Finally he will do so and following courtship his attack upon an aggressive stimulus is, in turn, held up for some minutes until his sexual arousal has declined. Rapid switches in behaviour of this type are unlikely to involve hormones. On the other hand as his reproductive cycle progresses the stickleback shows a sustained and increasing tendency to respond to parental stimuli from nest and eggs. This change almost certainly does have an endocrine basis (see Baggerman[11A]). One of the fastest hormonally-controlled responses is seen when frogs and chamaeleons change their colour to match their background. Visual stimuli lead to changes in the secretion of a pituitary hormone (**intermedin**) which, in turn, causes contraction or expansion of pigment cells of various colours in the skin. Frogs and chamaeleons begin to respond to background changes within a few minutes, but it may take half an hour or more to complete the colour change. This time scale can be compared with the second-to-second colour changes of an octopus, whose contractile pigment cells are under direct control of the nervous system.

THE VERTEBRATE ENDOCRINE SYSTEM

Because it provides the best examples, most of this chapter will be concerned with the reproductive behaviour of vertebrates and its endocrinological basis. As a preliminary, we must describe the endocrine organs and hormones which are of particular interest in this context. This account must be very brief—for more information consult Gorbman and Bern.[85] Our chief interest is with specific hormone effects, that is, those which influence the performance of particular types of behaviour. For this reason we shall not bother with such endocrine organs as the adrenal glands or thyroid glands. Their hormones affect behaviour but in a rather general way, the latter by means of its control of metabolic rate, for example. We have already considered in Chapter 5 some of the effects of adrenal hormones on behaviour in conflict situations.

The pituitary gland

Enough detail on the structure of the pituitary gland and its relationship to the hypothalamus has been given in Chapter 4. It secretes several hormones which affect the output of other endocrine organs and in this way the pituitary effectively controls the whole endocrine system. The hormones we shall be most concerned with are the **gonadotrophins** which act upon the gonads and promote both the growth of germ cells and the tissues of the gonads which secrete the sex hormones. There are two main gonadotrophins—**follicle stimulating hormone** (FSH) and **luteinizing hormone** (LH)—both were named after their action on the female gonads or ovaries, but they are secreted by, and have similar functions in males. In females, both are necessary for the growth of eggs and for their release into the oviduct ready for fertilization.

A third pituitary hormone important for behaviour is **prolactin**, also known as **lactogenic hormone** or **luteotrophic hormone** (LTH), which has a variety of physiological effects. We know that it is secreted by various classes of vertebrates but it often has completely different functions and 'target organs' (those parts of the body whose growth or functioning is affected by the hormone). In female mammals it stimulates growth of the mammary glands and milk secretion; it also stimulates the ovarian corpora lutea to secrete progesterone (see below). Prolactin was first discovered in birds where its most familiar effect is to cause the growth of the crop in pigeons. Both sexes feed their young on 'crop milk' which is formed by the breakdown of cells shed from the rapidly dividing crop epithelium. In some fish, prolactin controls colour changes by its influence on the pigment cells in the skin.

The gonads—ovary and testis

It has been known for centuries that castration has profound effects on the behaviour and body form of vertebrates. This is in contrast to many invertebrate groups in which castration has little or no outward effect. Only in the vertebrates are the gonads important endocrine organs where, under stimulation from pituitary FSH and LH, they produce the sex hormones from special secretory cells. The female hormones are collectively called *oestrogens* and the male hormones, ***androgens***. All these hormones are steroids which are closely related in chemical structure, and although different vertebrate groups secrete slightly different steroids, those from one group are usually quite effective in another. The commonest androgen secreted by mammals is called ***testosterone***.

The sex hormones are responsible for the development of the secondary sexual characters and for the growth of the reproductive system in preparation for the shedding of eggs and sperm. Usually there are permanent differences between the body form of male and female, which are maintained throughout adult life. These are often augmented by the seasonal growth of secondary sexual characters under the influence of increased sex hormones. Stags can be distinguished from hinds all the year round, but in addition they show seasonal growth of antlers.

Finally we must mention another steroid hormone produced by the ovary. After a mammalian egg has been shed, its empty follicle enlarges and forms a prominent yellowish structure on the ovary surface—the corpus luteum. This begins to secrete ***progesterone*** under whose influence the lining of the uterus is prepared for receiving the egg after fertilization and development into a blastocyst. Progesterone also inhibits the contraction of the uterine muscles, which must be avoided if pregnancy is to continue. It is justly called 'the hormone of pregnancy', but it is not an exclusively mammalian hormone. Structures like the corpora lutea form when eggs are shed in fish, amphibians and reptiles but we know little about the presence or action of progesterone in these classes. Birds have progesterone which is almost certainly secreted by the ovary although their corpora lutea are not conspicuous.

Interactions between hormones

A good proportion of the experimental work on hormones and behaviour involves the injection of hormones or implantation of hormone pellets beneath the skin from where they slowly diffuse into the bloodstream. Such experiments are essential, but their interpretation may need caution. One reason for this is that the presence of one hormone may be a trigger which starts secretion of a second. If we inject FSH into a reproductively inactive female and observe reproductive behaviour, we cannot ascribe this behaviour change to FSH itself. It is just as likely to be due to oestrogen, which FSH stimulates the ovaries to secrete. This hypothesis could be tested by injecting oestrogen alone. Again, the injection of a hormone may suppress the body's natural secretion of this same hormone. Circulating oestrogen feeds back upon the pituitary and suppresses FSH secretion, which in turn suppresses oestrogen secretion, so that the two hormones decline together. We are familiar with this type of effect from the action of oral contraceptives. Here 'the pill' taken daily artificially keeps up the level of oestrogen and progesterone in a woman's blood. This is sufficient to suppress the activity of her pituitary gland and, in particular, it no longer produces the pulse of LH secretion which would cause an egg to be shed from her ovaries.

Under natural conditions, more than one hormone may be in circulation at the same time. The combined effects of two hormones may be greater than either one alone and in such a case their action is said to be *synergistic*. Oestrogen and progesterone often act synergistically in female vertebrates. During nest building the body of the female canary is prepared for the incubation of eggs. A brood patch on her ventral surface becomes defeathered, highly vascularized and sensitive to touch: it is this patch that will cover the eggs as she incubates. Hinde and Steel[114] found that brood patch formation is under the influence of oestrogen, progesterone and prolactin. Oestrogen alone was effective but its action on defeathering and sensitivity was markedly enhanced when progesterone or prolactin was also injected. The two latter hormones had no effect unless combined with oestrogen.

The opposite to synergistic action is *antagonistic* action between hormones such as those involved in the reproductive cycle of some male birds, for example pigeons. At first they court the female and are aggressive towards other males; both actions influenced by testosterone. Later in the cycle they assist the female with incubation and rearing the young. Although we may regard progesterone as a female hormone because of its origin and effects in mammals it is in fact secreted in male birds—probably by the testis. It is this progesterone which acts antagonistically with testosterone and leads to the suppression of courtship and aggression in the male pigeon (see Hutchison,[124] and Erickson *et al.*[69D]).

Thus the vertebrate endocrine system operates by a complex system of interactions both positive and negative, with the concentration of one hormone affecting the secretion of others. In such a fashion hormone levels are naturally maintained in a delicate balance which large injections can easily disturb.

Endocrinologists often talk about 'a physiological dose' of a hormone, by which they mean one which produces roughly the same concentration in the bloodstream as does the natural source. Such doses are often very small indeed—a few millionths of a gram—and because hormones are effective in such minute amounts, larger doses can produce peculiar effects. Beach[23] has shown that massive doses of testosterone produce male behaviour from either male or female rats, but it also causes an increase in the amount of *female* sexual behaviour shown by both sexes. In such a case the hormone may be acting as some kind of general stimulant. Hinde and Steel[113] find that increasing the dose of oestrogen given to female canaries from 0·05 mg to 0·3 mg—which is certainly way above the physiological dose—actually *reduced* its effect on the development of sensitivity in the brood patch.

There is a further complication, which results from the close chemical relationship between the steroids, oestrogen, progesterone and testosterone. It is quite certain that the liver and other organs can convert one into another, particularly if there is a high concentration present in the blood. We know too that androgens and oestrogens are produced at several sites in the body apart from the gonads; for example the adrenals and, in mammals, the placenta once pregnancy is well established. Simple castration may not eliminate all sex hormone secretion.

With these complications in mind, we can now turn to discuss how hormones and nervous system combine to produce adaptive behaviour.

BEHAVIOUR AND THE ACTION OF HORMONES

There is now abundant evidence that hormones affect behaviour by direct action on specialized neurons in the central nervous system. However before we discuss some of this work, it is important to recognize that hormones are extremely potent chemicals which often produce rapid growth changes in their various target organs. Such changes may affect behaviour indirectly in ways that may be overlooked.

For example, in female mammals oestrogen causes growth of the uterus, changes in the epithelium of the vagina and growth of the mammary ducts. In birds oestrogen, sometimes acting synergistically with progesterone, causes enlargement of the oviduct and reproductive tract. In some species—as just described for the canary—it also causes the shedding of ventral feathers and increased vascularization of the ventral skin to form a brood patch. The mammary glands of a pregnant mammal and the crop of pigeons during the latter stages of incubation grow and become engorged under the influence of prolactin.

All these target organs have a nervous supply, and information on their growth may feed back to the nervous system and modify behaviour. Lehrman[147A] has shown that the engorgement of its crop with 'milk' is one of the factors that predisposes doves to feed their squabs. When he anaesthetized the nerves of the crop wall the doves' tendency to feed was reduced, presumably because they could no longer perceive crop distension.

Again, bodily changes produced by hormones may render an animal more responsive to certain types of stimuli. As mentioned in Chapter 4,

Beach and Levinson[25] have shown that testosterone causes a thinning of the skin covering the glans penis in male rats. This means that the tactile sense organs are more exposed and, under the influence of the hormone, the rats are made more sensitive to sexual stimulation. Similarly, Hinde and Steel[114] find that a female canary is more responsive to stimuli from the nest cup after its brood patch has developed, and this affects its behaviour during nest building and incubation.

Peripheral changes must not be ignored, but from the behavioural viewpoint the most important target organs for hormones are regions of the central nervous system itself. With improving techniques the study of the central action of hormones has advanced very rapidly over the past few years. As described in Chapter 4 it is possible to implant hollow needles into the brains of animals as small as a rat or a pigeon without interfering with their freedom to behave. Brain atlases of several species are now available from which, using stereo-taxic instruments, the tip of a needle can be placed within a fraction of a millimetre of the desired spot. The exact position can be checked later by sectioning the brain after behavioural tests are finished. Two basic techniques have been used. Firstly hormones in solution have been injected into neural tissue or into the ventricles or cavities in the brain; one of these—the third ventricle—is adjacent to the hypothalamus. Secondly, and more commonly, the needle before insertion has its tip coated with a waxy or crystalline form of the hormone (see Fig. 6.1). Refinements of this second technique have made it possible to control very precisely what surface area of hormone on the needle tip is in contact with the nervous tissue, and thus to control the rate at which it diffuses away from the tip.

In a now classic study of this type, Harris and Michael[98] investigated the rôle of oestrogen in the sexual behaviour of female cats. When on heat or in *oestrus*, as this receptive stage is called, female cats show a very characteristic posture which resembles in some respects that of the receptive female rat (Fig. 2.2). The rump is elevated, the tail deflected to one side and the hind legs making treading movements. A cat will assume this posture as soon as a male approaches, and will submit to being mounted. This is in sharp contrast to an unreceptive female—in *anoestrus*—who will lash out viciously if a male gets too close. Female cats normally show three oestrus cycles a year and each period of behavioural oestrus is preceded by slow changes in the reproductive system. The uterus grows and its walls thicken, the lining of the vagina thickens and begins to slough off cells. Only when these bodily changes are about complete, does the cat begin to show the behaviour typical of oestrus. It is possible to inject repeated small doses of oestrogen into the bloodstream of castrated females over a long period and produce complete growth of the reproductive tract without changing behaviour.[186] This immediately suggests that in cats the initiation

of oestrus behaviour does *not* depend on nervous feed-back from the enlarged reproductive system; i.e. the peripheral effects of oestrogen are not
involved in the behavioural changes.

Harris and Michael have produced clear confirmation of this by hormone implants. If a needle whose tip is coated with a slowly dissolving
oestrogen is placed in certain parts of the hypothalamus, castrated cats
often show strong oestrus behaviour. Most importantly they show full
oestrus behaviour whilst their reproductive system remains completely
undeveloped. The needle inevitably tears through a considerable volume of
brain tissue when it is inserted and clearly it is essential to have a series of

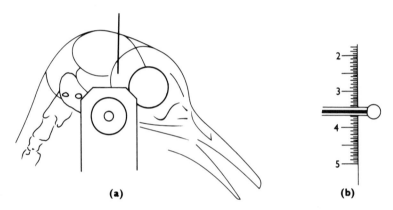

(a)   **(b)**

Fig. 6.1   (a) Lateral X-ray picture of the head of a dove showing the skull, held
fast in the clamps of a stereotaxic instrument, and with a fine tube carrying a
hormone implant penetrating down past the cerebral hemispheres into the anterior
hypothalamus; (b) shows a hormone implant fused on to the tip of fine steel
tubing. The large units on the scale are millimetres. (Drawn from photographs in
Hutchison,[124] 1970. *J. Reprod. Fert., Suppl.* **11**, 15.)

control experiments before it can be claimed that the observed effect is due
solely to oestrogen acting on the hypothalamus. Harris and Michael provide these. In one set of experiments they got oestrus behaviour in 13 out of
17 castrated cats with implants in the hypothalamus, but no case of oestrus
from 18 implants in other parts of the brain. Further, oestrus behaviour
was shown only when the needle was coated with oestrogen; other substances had no effect.

Towards the posterior part of the hypothalamus, oestrus can be evoked
from a fairly wide area, but further forward only those implants which
touch on a tiny area just dorsal to the optic chiasma (see Fig. 4.6) are
effective. This looks like an example of the phenomenon which we dis-

cussed in Chapter 4 where the same effects can be produced by stimulation at a number of different sites because they all impinge on the same converging or diverging fibre tracts. Certainly the initiation of sexual behaviour is relatively localized but it probably involves more than one group of neurons. These neurons appear to have a special affinity for oestrogen which they pick up from the bloodstream. Michael[185] found that when he injected radioactive oestrogens, whose exact position can be picked up later in sections of the brain, they were concentrated in just these same areas of the hypothalamus.

One of the problems posed by Harris and Michael's experiments concerns the delay between implanting oestrogens and the appearance of behavioural oestrus. This varied from 4 to 106 days, but once begun oestrus behaviour was just as persistent in cats with a long latency, often lasting for 6 or 8 weeks. Presumably the latency has something to do with rate of diffusion of the hormone away from the needle tip and into responsive neurons.

Recent studies with other animals have confirmed the general conclusions derived from the cat experiments. For example the work of Hutchison[124] and Komisaruk[140A] with implants in doves has revealed regions in the brain which are especially sensitive to hormones. The implantation of testosterone in the anterior hypothalamus and the pre-optic nucleus (so called because it lies just anterior to the point where the optic chiasma enters), causes castrated male doves to show courtship behaviour and aggressiveness. Progesterone implanted in the same places suppresses these behaviour patterns, and causes an increased tendency to incubate— a direct demonstration of the antagonism between progesterone and testosterone which we mentioned earlier.

There is also neat confirmation of the rôle of the hypothalamus in initiating not just courtship and copulation, but the whole pattern of reproductive behaviour. Thiessen and Yahr[252A] have worked with the Mongolian gerbil in which males mark their territories with an oily secretion from a gland on their ventral surface. Castration leads to regression of the gland and to the disappearance of marking behaviour. Testosterone implants—again in the pre-optic area already mentioned—or minute injections into the lateral ventricles cause castrated male gerbils to go through the full patterns of marking. Their ventral glands, however, remain completely undeveloped because not enough testosterone gets out into the blood stream. Thus there can be no doubt about the central control of marking behaviour.

Many problems remain concerning the central action of hormones on behaviour. For example there are anomalies concerning the latency of action. As mentioned, oestrogen implants in cats often take many days to produce their effects but oestrogen injections into the wing of muscles of doves produce their maximum effect upon behaviour within 30 minutes, (Vowles and Harwood[270A]). Before we can interpret such diverse results we shall need to know much more about the nature of the central nervous target organs and how the hormones get to them.

Further problems concern apparent anomalies in the specificity of hormone action. For example, Hutchison[124] found that he could elicit courtship behaviour (admittedly of low intensity) by oestrogen implants in male doves, whilst Fisher[72] elicited maternal behaviour from male rats by injections of testosterone into the anterior hypothalamus. It may be that the similar molecular structure of the various steroid hormones is the explanation here but, on the other hand, we may be wrong to expect complete specificity of hormone action. Such an expectation may imply that we are overemphasizing the specific organization of motivational systems and are thinking in terms of separate hormones affecting separate systems. Hormones may affect more than one system or they may act on more general brain mechanisms—for instance those affecting the animal's attention to external stimuli and the degree to which it persists in a response once initiated. Andrew[9A] has suggested that the action of testosterone in male vertebrates is of this type.

One of the most promising developments in the study of hormone action is the use of refined chemical assay methods to measure the hormone content of the blood directly. Already with larger animals one can take a harmless amount of blood—say 5 ml—at intervals and measure changes in circulating hormone levels. Soon it will be possible to analyse smaller samples and apply this technique to rats and mice. Direct hormone assay will help us to associate changes in the body's normal hormone secretion with changes in behaviour. Of course, the level of a hormone in the blood does not tell us accurately what levels are to be found in the central nervous target centres—we have seen that it is possible to change levels in one without affecting the other—but it will be a useful guide.

Hormone assay work in man has already yielded some fascinating results. Testosterone secretion in men shows a marked diurnal rhythm: it is highest in the morning and lowest around 8 p.m. (see Nieschlag and Ismail[199A]). It is tempting to speculate what effect this may have on male sexual tendencies, although one has little difficulty in concluding that other factors will also influence the timing of human sexual behaviour! Recently data has become available from volunteers who have taken samples of their own blood during arousal and following sexual intercourse. In every case the blood level of testosterone has been higher at such times than during the rest of the day, Loraine[154B] describes these results in more detail. This suggests that the actual performance of behaviour influences hormone secretion and such a conclusion is abundantly confirmed from other studies on animals.

THE INTERACTIONS BETWEEN HORMONES AND BEHAVIOUR

Hormones affect behaviour, but 'interaction' implies influences in both directions. Not only the quantity but also the type of hormone secreted can be affected by behavioural events.

We have already described some of the hormonal and behavioural changes in the oestrus cycle of the female cat. Fig. 6.2 summarizes and extends these to complete the picture. In most female mammals ovulation and oestrus behaviour are synchronized. The female becomes receptive to males, as a result of oestrogen acting on her hypothalamus, at about the same time as eggs are released into the oviducts following a 'pulse' of LH from the pituitary gland. It is likely that activity of the 'sex centre' in the hypothalamus triggers off the pituitary activity at the same time as it initiates oestrus behaviour. However, in the cat and a few other mammals there is an even more perfect synchronization, as indicated in Fig. 6.2. Cats do not ovulate until they have mated. The effects of copulation are conveyed by the nervous system to the hypothalamus where they initiate the pulse of LH from the pituitary. A few hours later ovulation occurs at just about the time when sperm have penetrated to the upper levels of the reproductive tract. This same sequence of events makes the rabbit—another 'induced ovulator' —an ideal subject for the mammalian embryologist. He knows that about 10 hours after mating freshly shed eggs will be found in the upper reproductive tract. Further, the effects of mating can easily be imitated by stimulating the vagina and cervix with a glass rod. It is interesting to note that an infertile mating is always followed by a period of 'pseudo-pregnancy' in which the female's body goes through all the normal changes for pregnancy. Female rabbits show enlarged mammary glands and loosening of the ventral hair; towards the period when birth would occur they even pluck the hair and build a nest. All these effects are the result of prolactin secretion which, in turn, stimulates the corpora lutea to produce progesterone.

Here then is a direct link between the behavioural act of mating and the pituitary gland which is stimulated, via the hypothalamus, to secrete LH and prolactin. In many mammals the initiation of an oestrus cycle by FSH secretion is primarily in response to changing day length. Female ferrets come on heat in early spring, no matter how cold it is; sheep come on heat in autumn. In the cat the initial FSH secretion seems to follow an inherent rhythm in the brain or pituitary gland whose exact nature is unknown. The same is true of rats and mice which usually show a regular 4-day oestrus cycle. The operation of this clock can be greatly affected by the behavioural situation. Whitten[281c] reviews a number of his own experiments showing that if female mice are kept together without males their oestrus cycles

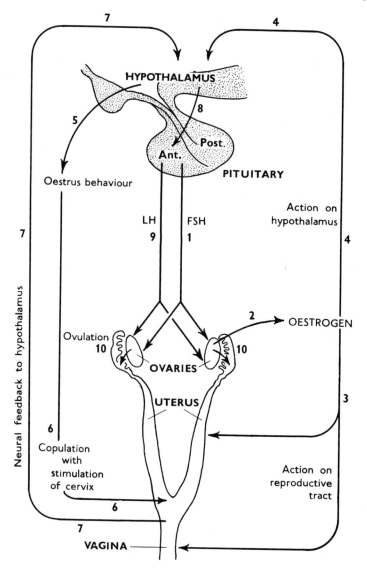

Fig. 6.2 A diagrammatic representation of some of the hormonal and behavioural events during the reproductive cycle of an 'induced ovulator' such as a cat or rabbit. The numbering of arrows indicates the approximate sequence of events—there is probably considerable overlapping in some cases.

become irregular and they may spend long periods in anoestrus. New cycles start the moment a male is put in with them, and the majority of such females come synchronously into oestrus on the third night after his introduction.

No such effect is produced if the male has been castrated, and his effectiveness is due to some substance excreted in his urine. In more recent experiments, Whitten *et al.*[281D] have proved that this substance is effective when airborne and females in no direct contact with males must detect it by smell. Such a substance, secreted to the exterior and acting like a kind of ectohormone is usually called a *pheromone*. Pheromones are not un-common, particularly in insects and mammals which are both groups with a well-developed olfactory sense (see reviews by Schneider,[233] Wilson,[283A] and Ralls[208A]). Both male and female pheromones are now recognized to be very important in mammalian reproduction and such knowledge can be put to good effect in animal husbandry. Sometimes a sterilized ram (i.e. one with the sperm duct severed, but otherwise intact) is allowed to run with the ewes before the breeding ram is introduced. The smell of the former ram helps to induce oestrus in the ewes and thereby improves fertility.

Work with the reproductive cycles of birds has also demonstrated a complex interaction between behaviour and hormone secretion. It has been known for more than 50 years that female pigeons do not normally lay eggs if kept alone or with other females, but begin to lay soon after a male is introduced. This must mean that pituitary secretion is influenced by the male, and simple tests show that the sight of a male is enough to cause ovulation, provided he is able to court the female. Thus a male who performs the typical courtship bows, even though separated behind a glass screen, is much more effective than a castrated, and therefore inactive, male in the same cage.

Lehrman and his co-workers have investigated the behavioural and endocrine basis of the reproductive cycle in the ring dove. When males and females are kept separately in large stock cages, they show few signs of reproductive activity. However, if a pair is put into a cage together courtship begins within a few minutes and this marks the beginning of a reproductive cycle. The birds court and mate, build a nest, lay eggs, take turns to incubate them, feed the squabs when they hatch until they leave the nest and can feed themselves, and then begin courtship again and the cycle repeats. This elaborate sequence occupies about 6 weeks and at each stage the bird's behaviour and physiological condition are both adapted to current needs and preparing for the next stage. The consideration we have given earlier on the time scale for neural and endocrine control certainly suggests that there must be underlying hormonal changes, and Lehrman's work reveals how these interact with external stimuli and the behaviour of the doves. We must condense the results of a whole series of elegant experiments; Lehrman[148-9] gives fuller accounts with references.

A male dove is usually ready to court the moment he is put with a female, which probably means that his testes are active and secreting testosterone. Visual stimuli from his courtship activate hypothalamic centres which control pituitary secretion in the female and induce the release of FSH and LH. These stimulate the growth of her ovary which in turn secretes oestrogen under whose influence her reproductive tract begins to grow. After a day or so both birds select a nest site and begin to build. The start of nest building may be related to increased secretion of sex hormones, but this point is still uncertain. During nest building courtship continues and the birds copulate. When the nest is complete the female becomes increasingly attached to it and she soon lays her first egg. This follows the release of LH from her pituitary, and for a few days previous to this her ovary has begun to secrete progesterone which induces incubation behaviour. The male, too, secretes progesterone which acts antagonistically with testosterone so that his courtship and aggression die away and he becomes willing to incubate eggs. The birds take turns to sit on the eggs and after a few days' incubation the first signs of growth in their crops can be seen which indicates that their pituitaries have begun to secrete prolactin. The act of incubation itself probably brings this about by feed-back to the hypothalamus which 'switches on' the pituitary. Prolactin not only leads to crop growth but also inhibits the secretion of FSH and LH which, in turn, shuts off the secretion of sex hormones and sexual behaviour between the pair dies down. After 14 days of incubation the eggs hatch and the squabs are fed on crop milk. This is pure at first but after a few days an increasing proportion of solid food appears in the milk which matches the growth of the squabs. After 10 to 12 days the young birds leave the nest, by which time parental feeding and prolactin secretion have begun to decline. As prolactin levels in the blood fall so FSH and LH can be secreted and the male soon starts to court the female again.

Some of the experiments involved in working out this sequence are quite dramatic. Thus the secretion of prolactin normally follows when the doves have been sitting on eggs for a few days, but a male dove's crop develops even if he does not actually incubate so long as he can see his mate incubating! Here then a specific visual stimulus leads to the secretion of a specific hormone, but it will do so only if the male is in a particular physiological state. He must previously have participated in nest building, because if he is separated from his mate earlier in the cycle his crop does not develop, even if he can see her incubating.

The general organization of the dove breeding cycle is typical of all birds, although the details of hormone action at the different stages vary. Thus in the dove progesterone secretion is certainly important for the start of incubation and prolactin is almost ineffective. In chickens, on the other hand, it is clear that prolactin alone can initiate incubation behaviour and produce a 'broody' hen.

The external stimuli which operate as cues in the reproductive cycle also vary from species to species. They are adapted to the particular circumstances which operate during the normal breeding of the species. Natural selection can provide some remarkable variations on the basic theme. Female cuckoos do not build nests but respond to the sight of their host's nesting behaviour. A cuckoo's ovulation is so timed that it can lay an egg in each of several nests just as the host bird completes its own clutch (Chance[48]).

In a few wading birds—the phalaropes and the dotterel—there is almost a complete reversal of the normal rôles of the sexes. It is the female who is brightly coloured, who fights with other females, courts the dull-coloured male and, having laid eggs in the nest which he has built, leaves him to incubate and raise the young. In phalaropes it appears that the hormones have retained their 'normal' function and females actually secrete more testosterone than do males. Johns[132] found that injecting males with testosterone causes them to produce the brightly coloured female plumage after a moult. Females also become aggressive just about the time that they develop their bright plumage in spring, and this is also the normal 'masculine' rôle of testosterone. Johns found that these effects of testosterone were not altered if he injected oestrogens at the same time. This would lead one to conclude that the female phalarope must secrete more testosterone than the male and recent work by Höhn[117B] has demonstrated that this is indeed the case. Interestingly enough some androgens are secreted by the ovaries of all birds and they may be a precursor for the synthesis of oestrogen. The ovary of the phalarope releases both testosterone, in large quantities, and oestrogen into the bloodstream, i.e. the female phalarope is endocrinologically bisexual.

Finally in this section, a word of caution. Not every adaptive sequence of reproductive behaviour necessarily has a hormonal basis. Each case must be approached with an open mind. Female mice show a nicely graded degree of responsiveness to their young. The newly-born young are given maximal attention and care, but as they become stronger the mother mouse leaves them more and more alone and begins to wean them. It is tempting to correlate this slow reduction in maternal responsiveness with the slow fall in prolactin secretion which we know takes place during weaning. However Noirot[200] has shown that whilst some hormonal influence may be operating, virgin females who have never seen young mice, show a similar level of maternal responses to young of various ages as do lactating females, though of course they cannot produce any milk. As the young mice grow they provide a changing stimulus and this determines how much maternal behaviour they elicit. Almost any adult mouse—including, in some cases, males—will respond appropriately.

HORMONES, SEXUAL DEVELOPMENT AND SEXUAL EXPERIENCE

Hitherto we have considered the action of hormones largely as agents which evoke from the animal behaviour patterns, the potential to perform which has already developed. However we know that hormones are just as important for determining how animals respond during the period when their inherited behavioural tendencies are developing and later, when they are being modified by experience.

Perhaps the most dramatic example of the action of hormones in development is their role in the determination of sexual responses. Here we must summarize briefly a complex story. It now appears that—at least in rodents—the hypothalamic mechanisms responsible for initiating sexual behaviour are 'labile' until the time of birth. This means that whichever the genetic sex of an embryo, the hypothalamus can be influenced to mediate male or female sexual behaviour in later life depending on one critical factor. This factor is the presence or absence of minute quantities of steroid hormones around the time of birth. If steroids *are* present the hypothalamus/pituitary gland complex tends to develop in a masculine fashion. In particular the pituitary gland secretes FSH and LH in a non-rhythmic way. If steroids are *not* present, then development tends to be of the female type and the pituitary gland secretes FSH and LH rythmically, so that typical oestrus cycles occur.

Levine and Mullins[150B, 151A] review the whole story which is inevitably rather less clear-cut than the above summary might suggest. Nevertheless it does appear that in normal development the testis of male embryos becomes active for a brief period around birth. The testosterone secreted is sufficient to set the young rodent on a masculine course of development. By contrast, the embryonic ovary does not secrete, the female hypothalmus is not stimulated and development is feminine. If a single, minute injection of testosterone is given to a female rat or mouse within three days of birth, it becomes acyclic, sterile and masculinized in its behaviour. Such females show a raised frequency of male sexual responses, mounting and thrusting, and they are also much more aggressive than normal (Bronson and Desjardins[38A]).

A similar treatment with oestrogen also has a masculinizing effect on females (though it is less powerful than testosterone) but in males oestrogen and testosterone do not work in the same direction. Oestrogen administered to a male at birth tends to act antagonistically with his own testosterone and he is partly feminized. However castration of a male at birth deprives him of his tiny pulse of testosterone and he is markedly feminized with a cyclic secretion of FSH and LH as in females.

Castration after ten days of age has no such effect, and its effects can be largely compensated by testosterone injections. Castration within the critical period during which the hypothalamic mechanisms are being determined has permanent effects, and no amount of testosterone can restore normal function.

These are very striking, all-or-nothing effects in which hormones are acting at a very early stage in the development of behaviour. In adult life we can study more flexible interactions between hormones and experience. The most direct way to approach the problem of adult development is to consider what happens when the normal hormone supply is cut off.

With the sex hormones, deprivation is most easily achieved by castration. The effects of castration on young vertebrates before they first reproduce are very uniform. Castrated animals, whether they be fish, amphibia, reptiles, birds or mammals, show no development of secondary sexual characters and rarely show any sexual behaviour. (There are a few exceptions to this rule amongst man and the higher primates.) The results are not so uniform if castration is performed later in life, after the animals have had sexual experience. Regression of sexual behaviour is normally complete in all the lower vertebrates but an interesting scale is revealed within the mammals.

Most workers report a rapid decline in sexual behaviour following castration in rats, mice and guinea-pigs. At the other extreme in man and the other primates, the castration of adults often has very little effect on behaviour. Intermediate are the dog and the cat, where sexual behaviour usually declines but does so slowly over a period of months or years and some individuals may show no decline. In general, throughout the mammals, females are more affected by castration than males. This scale is probably justified for the most part, but as McGill and Tucker[168B] have pointed out, we have not really been comparing equivalent samples of the different groups. The rats and, in particular, the mice have come from laboratory strains most of which are highly inbred and inbreeding is often accompanied by a general depression affecting both metabolism and behaviour. Conversely the dogs, cats and primates studied will all have come from outbred and very heterozygous stocks. McGill and Tucker did indeed find that castrated males from some heterozygous stocks of mice showed very slow decline in sexual responsiveness—some individuals continuing to mate for over a year, i.e. almost as persistently as cats and dogs.

Nevertheless, it is reasonable to construct some kind of scale within the mammals which, though not in any sense phylogenetic, does represent the partial emancipation of sexual behaviour from hormonal control and the increasing dominance of the brain's higher centres, particularly the cerebral hemispheres. The main control of the rat's sexual behaviour is vested in the hypothalamus, and damage to the cerebral hemispheres has little effect on mating. Cats and dogs are affected by such damage and in primates some cerebral lesions can abolish sexual behaviour altogether. We know that the cerebral hemispheres are involved in the storage of experience of all kinds. Consequently it is not surprising to find that the higher we proceed up this mammalian scale the more individual experience determines how easily the animal is aroused sexually and what stimuli are effective.

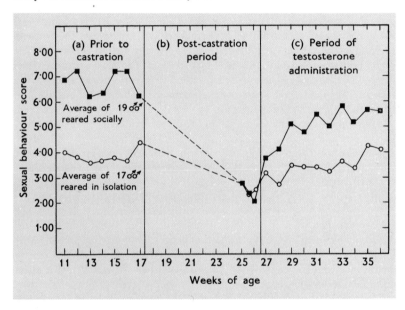

Fig. 6.3 Levels of sexual behaviour in male guinea-pigs reared socially and in isolation (a) before castration, (b) after castration, and (c) during artificial replacement of testosterone by injection. (Modified from Valenstein and Young,[267] 1955, *Endocrinology*, **56**, 173.)

However, it is important to emphasize that the sexual behaviour of all mammals is affected to some degree by their individual history and also that circulating sex hormones are essential if this behaviour is to develop normally. For example, Rosenblatt and Aronson[219] have shown that the degree to which a male cat's sexual behaviour persists after castration depends to a great extent on how much experience it had previously. Goldstein[83] has found a similar relationship in dogs. In neither species does any sexual behaviour develop if males are castrated before puberty. In other words, a male's experience can only affect its behaviour when sex hormone is circulating, but once stored in the brain the effects of experience persist in the absence of hormone.

Valenstein and Young[267] provide another example of the interaction between experience and hormones in behavioural development. Male guinea pigs inherit the basic patterns of sexual behaviour, but the frequency and persistence with which they are performed increases if they are reared with females after weaning. Valenstein and Young measured the sexual behaviour of males, half of which had been reared socially, the other half as isolated individuals. Figure 6.3a shows their sexual behaviour 'scores' on successive tests between 11 and 17 weeks of age. The 'score' is a compound one, which takes into account all the sexual behaviour patterns. The isolated animals scored consistently lower than the socially-reared animals. At 17 weeks both groups were castrated (Fig. 6.3b) and after 8 weeks further tests showed that sexual behaviour had declined to a very low level in all the animals. They could no longer be distinguished in behaviour because the effects of experience did not persist, as they do in cats and dogs. However, the difference was still there, latent, and was gradually revealed when Valenstein and Young replaced the natural sex hormones with daily injections of testosterone, Fig. 6.3c. The sex scores of both groups returned to their original levels and the difference between them was re-established.

HORMONES AND INSECT REPRODUCTIVE BEHAVIOUR

In conclusion, we may consider briefly some of the recent evidence from this rapidly growing field. It is important because it shows that animals with a completely different structure and evolution from the vertebrates nevertheless use similar methods to co-ordinate their behaviour with their reproductive physiology.

The rôle of hormones in the growth and moulting cycles of insects has been familiar for some years following the classic investigations of Wigglesworth[282] and others. It is only recently that experiments have shown how the same endocrine organs and probably the same hormones play a rôle in the control of adult behaviour. Most of the work has been on cockroaches and locusts and is reviewed by Engelmann.[69C]

Fig. 6.4 is a diagram of the main endocrine organs which affect the behaviour of insects. There is a pair of glands (which sometimes fuse together) called the ***corpora allata***, whose hormone promotes growth of the ovaries in adult female insects. As with the vertebrate pituitary gland, the activity of the corpora allata is under the control of the nervous system. Nerves may supply them in the usual way, but the allata are also controlled by modified neurons (the neurosecretory cells) in the protocerebrum of the brain. These pass their secretion along their axons to the allata.

The gonads of insects do not themselves secrete hormones, so the synchronization of sexual responsiveness with maturation of the ovary or testis cannot follow the same course as in the vertebrates. Instead the hormone of the corpora allata or of the neurosecretory cells may control the responsiveness of females. In most insects so far analysed, the hormone operates only by determining how attractive females are to males. Most female cockroaches whose ovaries are mature, but which have not yet mated, secrete

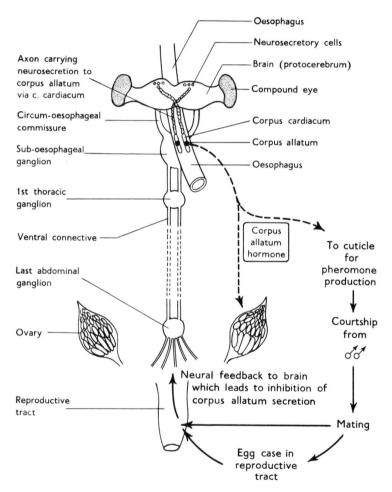

Fig. 6.4 The principal endocrine organs of a female cockroach as they affect reproductive behaviour.

a pheromone from their body surface whose presence stimulates males to court them. Males are aroused even if they just touch a piece of filter paper which has been in contact with pheromone. Corpus allatum hormone is needed before a female will secrete pheromone and both hormone and pheromone production stop after a female has mated. Once the first batch of eggs has been laid, the corpora allata become active again, more eggs mature, more pheromone is secreted and the female will mate again. Virgin females whose corpora allata have been removed do not produce pheromone and are therefore ignored by males. If pheromone is rubbed onto their bodies some of them respond when courted and mate. However some females do not respond normally and it appears that the corpus allatum hormone also affects their receptivity to males. Most of these experiments have used the cockroach *Leucophaea maderae*, but there is considerable variation between different cockroach species. Roth and Barth[221] have suggested that the behavioural receptivity of females in some species is controlled by neurosecretory hormone acting directly on some brain mechanism. Barth[16A] provides a full review of recent work.

In some grasshoppers—quite close relatives of cockroaches—the behaviour of females is under direct control of corpus allatum hormone. Loher and Huber[154A] have shown that females of the grasshopper, *Gomphocerus rufus*, refuse to accept males if their corpora allata have been removed even though, because pheromones are not important, they are courted well. Their receptivity returns about a week after glands from another insect are implanted. The ovaries themselves cannot be influencing this because their removal from young females does not affect behaviour.

In the fruit-fly, *Drosophila*, there is a similar control of receptivity. When females first emerge from the pupa their ovaries and corpus allatum are both small. The gland enlarges and its hormone causes rapid growth of the ovaries; ripe eggs are found after 48 hours or less. Very young females will not accept males although they are courted persistently. At about 36 hours from emergence they quite suddenly become receptive although they still require, on average, 2 to 3 minutes of courtship before they accept a male. Females kept without protein on a diet of sugar alone, show scarcely any ovarial growth but they go through the same receptivity cycle as normal. As with *Gomphocerus*, receptivity is 'switched on' when the concentration of corpus allatum hormone reaches a critical level, normally attained just before the female is ready to lay eggs. Injecting a mature corpus allatum into pupae causes the females which emerge to become receptive earlier than normal.[178]

Gwadz[91A] has found two distinct behavioural types in different populations of a mosquito (*Aëdes atropalpus*) which lives in rockpools. In one type the females become receptive soon after emergence and are mated within two days. The other type develops receptivity much more slowly, mating after about five days. Gwadz could show that receptivity depended, as with *Drosophila*, on the secretion of corpus allatum hormone, which occurred much later in the slow strain. This difference, whose function is obscure, depends on a single gene. It provides a clear example of one way genetic differences can affect behaviour; in this case through a change mediated by the endocrine system. We must now turn to discuss genetics and how it is involved in the evolution of behaviour.

7

Evolution

All through this book we have been using a biological approach which emphasizes the adaptive rôle of behaviour in an animal's life. The concepts of evolution have been implicit and often explicit, and consequently there is bound to be some overlap between this chapter and the others.

Here we shall concentrate on some selected aspects of behavioural evolution—in particular its relationship with genetics—and also on some of the ways in which behaviour can influence the course of evolution. By their behaviour, even the simplest animals can change their environment to suit themselves and thus modify the selective forces which affect them. Wood-lice move out of dry places into moist ones where they survive better; blue tits select an area of deciduous woodland for breeding because it provides the most food, and so on.

Of course animals do not simply rely on choosing from a range of environments that are already available; many can modify external conditions to match their requirements. Beavers dam up streams to provide deeper water; termites avoid desiccation whilst foraging by constructing covered paths over dry terrain; both termites and honey bees use a complex range of stereotyped behaviour patterns to construct around themselves a nest or hive whose physical characteristics—temperature, light and humidity for example—are kept rigidly controlled.

We are familiar with the overall adaptiveness of animal behaviour and the way in which it is correlated with morphology, but it often requires detailed study to reveal just how perfect such adaptation is. For example, there are two or more species of mole cricket (*Gryllotalpa*) living in Europe.

* A fuller treatment of many topics covered in this chapter can be found in Manning.[178B]

They are large burrowing insects and the males sing from their burrows below ground. As with many crickets and grasshoppers the song of each species is highly distinctive; (they certainly play a rôle in the sexual isolation between species, see p. 167). The character of the song is determined in part by the structure and size of the forewings, which are rubbed rapidly together to produce sound. A scraper or plectrum on the hind margin of the left wing rubs along the toothed under surface of a vein on the right wing (the 'file'). In two species studied by Bennet-Clark[26A, 26B] the fundamental sound frequency within the pulses of the song (see Fig. 7.1) was about 1,600 Hz in *G. gryllotalpa* a species with small wings, shallow teeth on the file and a quiet song, but 3,500 Hz in *G. vineae* which has large wings, deep teeth on its file and a much louder song.

Here then we can see the direct correlation between behaviour and morphology, but the mole cricket's adaptiveness goes much further than this. The male of each species excavates its burrow in a different way, illustrated in Fig. 7.1. Bennet-Clark has shown that the burrow forms an

Fig. 7.1 Side views of males of two species of mole cricket (*Gryllotalpa*) sitting in their burrows, head downwards in the singing position. The shape of their burrows approximates to an exponential horn. *G. vineae*, on the right, has a smoother tunnel and a much louder song, but in both cases the shape of the burrow is adapted to their song frequency (oscillograms of which are illustrated below) so as to emit it with maximum efficiency. (Adapted from Bennet-Clark,[26B] 1970. *Proc. R. ent. Soc. Lond.* (B), **39**, 125–32.)

exponential horn with a bulb at its base, whose physical properties are matched to the frequency of the song, emitting it with the maximum efficiency. When singing the male positions himself head downwards so that sound is produced just at the origin of the horn, as shown in Fig. 7.1. Thus the wing structure, singing and digging behaviour of the mole crickets are all precisely co-adapted each with the others, so as to provide the most efficient sound production, 'aimed' at the females who are attracted to the male burrows as they fly above.

As a further example of the precision with which behaviour has evolved to meet the demands of the animal's environment, we may refer again to the habit of black-headed gulls removing the empty eggshells from their nest. (The stimuli to which they respond in this situation were described on p. 47). Gulls breeding for the first time carry away eggshells promptly although they never pick up eggshells in any other situation. It is an activity which occupies them for about 5 minutes each year and might, for this reason, seem of trivial importance. Nevertheless Tinbergen and his collaborators[263B] have clearly shown that eggshell removal is important for camouflage of the nest. Predation of chicks by crows, stoats and foxes is rife in gull colonies and such predators discover nests and chicks more easily if the shells with their smell and white shell membranes (see Fig. 3.4) are nearby. Natural selection will have operated to ensure that parent gulls never fail to perform this brief but vitally important behaviour. Clearly we must never write off as functionless—however trivial it may seem—any piece of behaviour which we observe regularly in the natural situation.

In Chapter 2 we discussed development and the manner in which both genetic and environmental factors contributed to the final form which behaviour takes in the adult animal. Behaviour can evolve only if it varies and if such variations are transmitted from one generation to the next. Then natural selection can come into operation, favouring some variants—which thus spread—and eliminating others. With most morphological and physiological characters there is only one way in which such inherited variations can be produced—by genetic changes which will involve new mutations or recombinations of genes already present.

Behavioural evolution has another dimension because an animal may learn certain patterns of behaviour from its parents or others in its group. Later it may in turn act as a model from which its own offspring will modify their behaviour. Suppose now one animal learns a completely new behaviour pattern which proves more successful for some purpose than the hitherto typical behaviour. This pattern may be passed on to succeeding generations, and gradually replace the old pattern, without any genetic changes being involved.

We can begin our discussion of evolution by considering some examples of this relatively rare, but potentially very important, manner of behavioural transmission in animals. We are familiar with such 'cultural' evolution in human behaviour, and many would argue that almost everything of importance in human behaviour is transmitted in this way from one generation of a society to the next. The various human languages are an obvious example of a continuous cultural tradition which maintains different types of behaviour in different populations.

Clearly cultural evolution is only possible among animals with a considerable ability to modify their behaviour by copying and practice. We might not be surprised to find examples from the other primates, our closest relatives. Intensive, long-term observations on the Japanese monkey, *Macaca fuscata*, has shown that some differences between the behaviour of monkey troops are indeed of cultural origin. The monkeys are attracted to feed on corn, sweet potatoes and other foods artificially put down for them. The potatoes are often caked with earth which monkeys usually rub off with their hands. One day observers saw a young female dip her potato into a stream and wash it. She persisted in this habit, which was copied, first by one of her infants, and subsequently by nearly all the younger members of her troop, which is now clearly to be distinguished as a 'washing sub-culture'. A wide range of feeding differences between troops have become established in a similar manner (Frisch[76A] reviews some of the Japanese work). There seems little doubt that close observation of other primate groups will reveal a similar story (see Chapter 10).

We can readily accept cultural transmission in primates, but there is growing evidence that it occurs, though perhaps sporadically, in other vertebrates. It is cultural transmission that maintains the different song dialects in white-crowned sparrows as described on p. 33. The young male models the finer details of his song on that of his father and neighbouring males singing close by. We know that he would pick up a different dialect if he grew up in another area.

Recently Norton-Griffiths[201B, 201C] has discovered a remarkable case of cultural transmission in the feeding habits of the oyster-catcher, *Haematopus ostralegus*. These shore birds feed extensively on mussels and they have two distinct methods of opening the shells. One is by hammering on the ventral surface of mussels which are carried from the rocks to an area of firm sand suitable for the purpose. The second method involves stabbing the bill into the gape of slightly open mussels which are still attached to rocks and covered by sea water. Norton-Griffiths found that individual oyster-catchers use only one method to open shells. Young birds follow their parents down to the mussel beds and gradually acquire the same technique, which they use exclusively thereafter. Different mussel beds vary in the degree to which they are suitable for hammering or stabbing and oyster-catchers select areas which suit their own technique. There is certainly the basis here for real 'cultural isolation' between hammering and stabbing birds but, as Norton-Griffiths points out, oyster-catchers have many other foods besides mussels. Cultural transmission of novel feeding habits in birds has also been described for the familiar plundering of milk bottles by titmice,[112C] and for the way greenfinches (*Chloris chloris*) acquired the habit of extracting seeds from the unripe fruits of *Daphne mezereum*, a plant whose ripe fruits have inaccessible seeds.[206A]

However striking such examples are, we must recognize that they can only have a very limited rôle to play in the evolution of animal behaviour. As a rule this can only proceed if there are inherited variations upon which selection can operate. Consequently before we can profitably discuss what types of change have occurred during evolution, it is necessary to consider what is known about the genetics of behaviour.

GENES AND BEHAVIOUR

All behaviour has an inherited basis, but strictly speaking it is only a potentiality which is inherited. An animal will respond in a certain way to a certain range of stimuli given a certain set of genes and a certain range of environments during development. It is obviously going to be easiest to study the inheritance of behaviour in which the action of genes is well buffered against environmental changes. The stereotyped 'fixed action patterns' which are so conspicuous in arthropod behaviour and in the threat and courtship displays of many vertebrates are the best 'material' of this type. These patterns are so little affected by learning or practice that it is possible to study the effects of genetic changes against a relatively consistent background.

Learning ability is just as much under genetical control, but the effects of genetic changes are much harder to study. At present we can talk about the evolution of learning only in the most general terms and this fits best into the next chapter. Here we shall concentrate upon the genetics and evolution of fixed action patterns.

Behaviour involves the co-ordinated control of a number of different systems of the body. As a result genes which affect a whole variety of morphological or physiological characters can affect behaviour too. This is well illustrated from work on the fruit-fly, *Drosophila*, which is better known genetically than any other animal. Many hundreds of mutant genes are known and their position on the chromosomes has been mapped. For example, the gene called *Bar* reduces the number of facets in the compound eyes, *white* reduces pigmentation in the eyes, *forked* and *hairy* affect the number and structure of the bristles, *vestigial* and *dumpy* alter the form of the wings, *yellow* and *black* affect the general pigmentation of the body, and so on.

All these genes have been given names which roughly describe the most conspicuous effect they produce. They certainly have other effects too, which are sometimes detected in tests of metabolism, fertility, longevity and the like. Most of the genes mentioned above have also been shown to alter at least one aspect of behaviour—males which carry them are less suc-

cessful than normal males in stimulating females to mate with them. However, although this superficial description lumps all of them into the same
behavioural category, it is obvious that they produce their common end
result in quite different ways. *Bar* and *white* affect vision: mutant flies
cannot see as well as normals and will have trouble both in locating females
and in receiving visual stimulation from them during courtship. Bristles
are tactile sense organs, and males carrying *forked* or *hairy* genes will be
defective in this respect. An important part of the normal *Drosophila* male's
courtship is wing vibration in which he brings out one wing laterally and
vibrates it in the horizontal plane. This stimulates sense organs at the base
of the female's antennae. *Vestigial* and *dumpy* males have grossly malformed
wings which cannot be vibrated properly and it is not surprising that they
have poor mating success. It is less obvious why *yellow* and *black* males
take longer to mate than normal males. It is not their abnormal colour that
puts the females off, because they are equally at a disadvantage when
competing for mates in total darkness. There are no obvious deficien-
cies in the sense organs or wings, though the elaborate tests needed
to prove this have not yet been made. The genes could be operating
on the nervous system, the muscles or the general metabolism of the
insects.

It is not usually possible to identify the pathway by which a gene affects
behaviour unless the effect is large. The genes for microcephaly and phenyl-
ketonuria in man produce gross defects in the structure and biochemistry
of the brain with resulting severe mental defect. In mice there are a number
of genes known which upset their posture and balance. The 'waltzing
mice' spin round in circles when disturbed, others show continuous head
shaking or circling. These behavioural results can be correlated with
structural defects in the middle ear and brain centres responsible for
balance. Such changes are both large and deleterious, and on both counts
they have little relation to the kinds of genetic change—presumably small
and favourable—which have been important in the evolution of behaviour.

A less drastic change and a more precise analysis has been provided by
the remarkable work of Ikeda and Kaplan.[128A,128B] They have studied a
number of *Drosophila* mutants called collectively 'hyperkinetics', because
the affected flies show frequent leg twitches and tend to make sudden
jerks in their walking. Using refined neurophysiological techniques they
have located groups of motor neurons in the lateral part of the thoracic
neural complex which have abnormally high rates of spontaneous discharge.
Sudden bursts of firing from these neurons can be completely independent
of any external inputs: it really looks as if Ikeda and Kaplan have located
the neural sites of gene action: presumably the membrane physiology of the
affected cells is changed to produce the abnormal rates of discharge. In
this case we really can explain the behavioural changes—albeit still fairly

gross ones—in terms of the neurophysiology of gene action. Usually we can analyse the effects of such genes only in terms of the behavioural change which they produce. The methods used have, as far as possible, been those of classical genetics. Crosses are set up between animals whose behaviour differs and the effects are observed in the F_1 hybrids, in the F_2, in back-crosses and so on.

One immediate difficulty is the choice of suitable 'units of behaviour' for genetic analysis. This is one example of a problem which has always faced geneticists. Mendel's stroke of genius was to choose characters in his pea plants which were clear-cut and could be counted. In this case, each 'unit' —tallness, seed colour, texture of seed coat, etc.—corresponded to a single gene, but there is little probability of getting such a simple correlation with behaviour.

It is impossible to lay down rules for the selection of behaviour units; the genetical tests themselves will usually suggest the best ones to use. Fixed action patterns will normally depend on the operation of many different genes for their development, but simpler forms of inheritance may control the frequency with which such patterns are performed or the ease with which they are elicited.

Rothenbuhler's[222-3] elegant work with honey-bees is one of the best examples. Certain strains of bee are called 'hygienic' because, if a larva dies inside a cell, the workers uncap the cell and remove the corpse; 'unhygienic' strains do not respond in this way and the workers leave the cells alone. The hybrids between these strains are all unhygienic, which thus appears to be a dominant character. Rothenbuhler back-crossed the hybrids to the recessive, hygienic strain with the following remarkable result. From 29 back-cross colonies he found:

1. Nine of the colonies uncapped cells containing dead larvae, but left the corpses untouched.
2. Six did not uncap cells, but would remove dead larvae from cells which were uncapped by the experimenter.
3. Eight neither uncapped cells nor removed larvae, i.e. were unhygienic.
4. Six uncapped cells and removed larvae, i.e. were hygienic.

The proportions of the four classes do not differ significantly from equality and it is possible to explain this result in terms of two pairs of alleles, one of which controls the expression of the uncapping pattern, the other that of removing a dead larva. The 'unhygienic' alleles are dominant, we may call them 'U' for uncapping and 'R' for removing; then their 'hygienic' alleles are 'u' and 'r'. Worker bees of the hygienic strain must have the genetic constitution uurr, unhygienic workers have UURR. All the F_1 hybrids will have UuRr and be unhygienic, but back-crossing them to the recessive, hygienic strain gives:

The simplicity of this result does not imply that a single locus controls the development of whatever complex neuronal network organizes the fixed action pattern of uncapping or that of removing corpses. Rothenbuhler has evidence that unhygienic workers do perform these activities but at a very low frequency and they require a more powerful stimulus. The alleles U and u act as 'switches' determining the threshold of the uncapping pattern in a virtually all-or-nothing fashion.

We have other examples of fixed action patterns which inherit completely or not at all. The courtship display of ducks consists of a series of patterns most of which can be observed, with slight modifications, throughout the family. One such pattern Lorenz[156] calls the 'down-up' in which a drake dips his bill into the water and then suddenly lifts his head, raising a plume of water. This pattern is not seen in the yellow-billed teal (*Nettion flavirostre*) or the pintail (*Dafila acuta*), although found in a number of their relatives. However, the 'down-up' reappears in its typical form in the F_1 hybrids between these two species.[159,271] The most likely explanation for this is that the block of genes necessary for the organization of the 'down-up' are still present in both parent species but natural selection has eliminated its performance by raising its threshold. This threshold change has been accomplished by a different set of genes in each species. When they combine in the F_1 hybrids, their effect is reduced so that the threshold is lowered and the down-up is once more performed in the normal situation. We should need to study F_2 hybrids and back-crosses to confirm this hypothesis, but unfortunately, as is usually the case, the hybrids are very infertile.

We have described two examples where genes affecting thresholds may be operating rather like a 'switch' determining whether or not a pattern appears. 'Switch genes' are known in other circumstances,[240] controlling the operation of a group of genes which inherit as a block. Natural selection will favour linking together genes which all contribute to a common feature which must appear fully formed if it is to be of any use. The closer such genes are linked the less the danger that individuals will lack some of them and develop imperfectly.

We have a possible example of this from the work of Ewing[71A] who found that the specific form of the courtship 'song' (produced by the male's wing vibration display) in *Drosophila pseudoobscura* and *D. persimilis* is determined by the X chromosome. F_1 hybrid males produce 'songs' which resemble that of their mother's species since they receive their X chromosome from her. A number of genes must be involved in determining the song characters, and the fact that they are all linked together on the X chromosome can hardly be fortuitous. It is tempting to speculate that such gene blocks may be a feature of the inheritance of fixed action patterns. We need many more such studies, but it is not easy to find suitable material for genetic work of this type.

It is most interesting that when a number of fixed action patterns normally occur together in a particular sequence, often it is the sequence which becomes disrupted in hybrids, although the patterns themselves remain intact. The extraordinary nest building behaviour of hybrid love-birds described on p. 32, is a case in point. Lorenz[158A] and Ramsay[209] have described striking examples from the courtship behaviour of ducks. The drake mallard (*Anas platyrhynchos*), in common with other surface-feeding ducks, has a repertoire of some ten highly stereotyped courtship patterns some of which are illustrated in Fig. 7.2. The sequence shown—'bill-shake, grunt-whistle, tail-shake'—is called by Lorenz an 'obligatory sequence' although Ramsay mentions that occasionally he observed 'tail-shake, bill-shake, grunt-whistle' both in the mallard and in the closely related black duck (*Anas rubripres*). In hybrid drakes between these two species, he saw 'grunt-whistle, tail-shake, bill-shake', and other combinations which were never observed in either parent species. Clearly the genetic control of sequences is independent of that for the fixed action patterns themselves, but we know too little of the behavioural mechanisms involved to be able to comment further.

Whilst the 'switch genes' discussed above produce dramatic changes to the thresholds of behaviour patterns, it is far more common to find genes producing relatively small quantitative changes. We have mentioned the reduced mating success of male *Drosophila melanogaster* which carry the gene *yellow*. Bastock[17] has shown that their courtship is less stimulating to females because it contains a smaller proportion of wing vibration. Yellow males perform this fixed action pattern in the same way as normal males, but do so less often. A number of other examples of this type are known from *Drosophila* where it is easy to produce stocks which, as near as possible, differ by only a single gene.[176]

In other animals most studies have used either inbred lines, which will differ by many genes, or lines which have been deliberately selected for differences in behaviour. The same kinds of quantitative differences are recorded with the same fixed action patterns being performed at different frequencies. Thus different inbred strains of mice[168] and guinea pigs[86] show variations of this type in their sexual behaviour. Scott and Fuller[236] give numerous examples from domestic dogs, where the different breeds are the result of a combination of selection and inbreeding. It is a remarkable

'Bill—shake'   'Grunt—whistle'   'Tail—shake'

Fig. 7.2 One common sequence of stereotyped fixed action patterns in the courtship of mallard drakes. Hybrid drakes show these typical patterns but their sequence is abnormal (see text). (From Lorenz, K.,[158A]. 'The Evolution of Behavior'. *Scientific American.* © 1958 by Scientific American. All rights reserved.)

testimony to the stability of fixed action patterns, that some 8,000 years of domestication with, at times, intense selection, has produced so little modification of the ancestral patterns. All the behaviour repertoire of the domestic dog breeds is represented in the wolf but frequencies vary. For example, the terrier breeds bark frequently, spaniels do so less often, and wolves and basenjis scarcely ever bark. Similar frequency differences are found for all the other patterns.

Deliberate selection for behavioural characters has nearly always been successful in changing them in a quantitative fashion. For example, within three generations Wood-Gush[284] was able to change the frequency of sexual behaviour patterns in chickens to a marked extent. Manning[175] selectively bred *Drosophila* for fast and slow speed of mating over a number of generations; see Fig. 7.3. 'Mating speed' is a complex character which involves the interaction of male and female, but analysis showed that the behaviour of both sexes had been altered in a quantitative way. The males from fast-mating lines perform high intensity courtship movements more frequently than those from slow lines. Conversely, females from fast lines are more easily stimulated to accept males, of their own or from other lines, than are the slow mating females.

As well as changing the frequency of fixed action patterns, selection has been equally successful for altering more 'general' aspects of behaviour, such as locomotor activity and 'emotionality'. For example, Rundquist[225] selectively bred two lines of rats which showed respectively high and low activity in a running wheel attached to their home cages, which they could enter at will. He could produce little change in the direction of increased activity, but after 12 generations rats of his low activity line were averaging 6,000 turns of the wheel in a 15-day period as compared with over 100,000 for normal rats. Though we have no details, a change of this type will probably have effects on many different aspects of the rats' behaviour. So do changes to 'emotionality' which have been extensively investigated by Broadhurst.[37] He has selectively bred rats which show 'emotional' and 'non-emotional' (or as he terms it, 'reactive' and 'non-reactive') responses to a mildly frightening experience—being placed in an arena under bright light. We have already mentioned this type of 'open field' test when discussing the effects of early experience on emotionality (see p. 24). 'Reactive' rats are defined as those which show a lot of defaecation and urination and stay still, often crouching in one place. These animals are likely to be those which have a high degree of physiological arousal (see p. 97) as compared with the 'non-reactive' rats which do not defaecate and move around more in the arena. Broadhurst obtained a large response to selection, and the reactive and non-reactive lines have proved to differ in several behavioural and physiological measures. For example, reactive rats are slower to learn a simple response in order to avoid an electric shock[38]—their intense arousal 'interferes' with their learning (see p. 189).

The success of selection experiments indicates that natural populations carry numerous genes which affect behaviour quantitatively. Natural selection will maintain the optimum level of response, but there is plenty of variability for artificial selection to act upon.

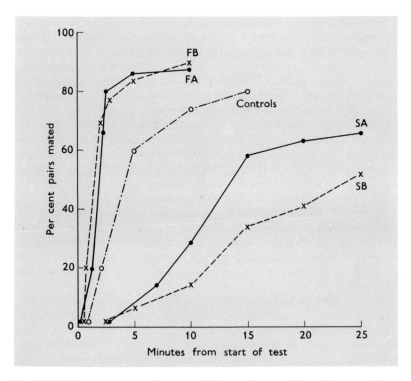

Fig. 7.3 The mating speed of groups of 50 pairs of *Drosophila melanogaster* from two lines selected for fast mating, FA and FB, and two selected for slow mating, SA and SB, compared with unselected controls. These samples are from the 18th selected generation. Some 80% of the fast lines have mated before the first of the slow lines begin. (From Manning,[175] 1961, *Anim. Behav.*, **9**, 82.)

Fuller and Thompson[78] provide a full review of work on the genetics of behaviour, together with some account of the theory and practice of genetics. Hirsch[116] contains a number of excellent reviews which include the more recent literature. Smith[246] includes an admirably clear account of the genetical basis of selection and evolution.

THE MICROEVOLUTION OF BEHAVIOUR

The term 'microevolution' is used to describe the earliest stages of divergence between populations leading up to the origin of species themselves. It is relatively easy to relate the behavioural changes involved in the microevolution of a group to the genetic changes we have just been considering.

If the behaviour of a group of related species is compared, it is usually possible to identify the same fixed action patterns throughout. These patterns are truly homologous, just as the skeletal elements of the different vertebrate groups are homologous—they have descended to each species from a common ancestor. However, the patterns themselves have been modified by selection in a number of ways. The commonest types of change affect (a) the frequency of performance of patterns, and (b) their form or 'emphasis'. The former category needs little further comment, it is exactly comparable to the types of inter-strain differences we have just been describing. To give just three examples from many, the ducks have a common repertoire of courtship patterns which appear at different frequencies in each species. Some, as we have noted, do not perform certain patterns though carrying the genetic potentiality to do so. In *Drosophila*, *D. melanogaster* and *D. simulans* show similar differences; in the former the commonest courtship pattern is vibration, in the latter, 'scissoring'—a pattern involving synchronized movements of both wings.[174] Clark *et al.*[51] have described frequency differences between homologous courtship patterns in two tropical fresh-water fish, the swordtail and the platyfish.

Microevolutionary changes to the form of patterns are equally universal; they produce different kinds of 'emphasis' to common display patterns. Two examples can illustrate this, more are given by Manning.[176] Figure 7.4 illustrates a typical case; the 'long call' of gulls is equivalent to the territorial singing of passerine birds. The two species shown go through the same sequence of head movements when calling but 'emphasize' different parts of the sequence and this gives a characteristic 'look' to each display. Crane[55] has made a thorough comparative study of the courtship displays of the fiddler-crabs (*Uca*). The males dig burrows on the beach and stand at the entrance; one of their claws is much larger than the other and is brightly coloured. When a female comes near they rhythmically wave this claw and, if the female is receptive, she eventually approaches. The exact form of the male's wave is distinct for each species, but the genus *Uca* can be divided into two main sub-genera on both behavioural and morphological criteria. The one, which Crane calls the 'vertical wavers', extend their claw only slightly, and merely lift it up and forward. The 'lateral wavers' extend the claw out sideways and lower it to complete a more or less circular path.

For convenience, we have isolated two types of change—frequency of performance and form of pattern—but clearly they normally occur together and combined with other types of change. For example, a great variety of changes are involved in the radiation of different types of cricket song from a common ancestral type—Alexander[2] discusses this with

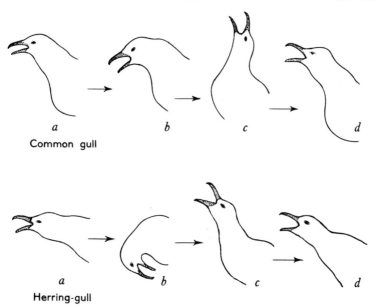

Common gull

Herring-gull

Fig. 7.4 Variations of emphasis in the 'oblique long call' sequence of two gull species. The sequence reads from left to right. At the start the head is in the oblique posture (a), it is jerked down as the bird begins calling (b), and is then thrown back (c) as the calls continue. The head is gradually lowered (d) as the calls die away. The common gull shows little emphasis of (b) and much of (c) ; the converse is true of the herring-gull. (From Tinbergen,[263] 1959, *Behaviour*, **15**, 1.)

numerous examples. Cricket songs vary in the amplitude, length and frequency of pulses and thereby produce a wide range of trills and chirps, each of which is quite distinctive. The same kind of factors must have been involved in the evolution of bird song. The song thrush (*Turdus ericetorum*) characteristically repeats each phrase of its song two or three times before passing on to the next.* Its close relative, the blackbird (*T. merula*), does not do so.

* 'That's the wise thrush; he sings each song twice over,
  Lest you should think he never could recapture
  The first fine careless rapture!'
                              Robert Browning.

THE NATURE OF GENETIC AND MICROEVOLUTIONARY CHANGES

It is worth while trying to discover if there is a common thread running through the diversity of genetic and microevolutionary changes just described. We have emphasized the quantitative nature of the changes and many of them can, in behavioural terms, be directly related to changes in thresholds. If, in the same situation, two species or strains differ in the relative frequency with which they perform patterns A and B from a common repertoire, it is reasonable to interpret this as due to threshold differences in the mechanisms controlling A and B in each case. There are numerous sites at which the control mechanisms could be affected. There might be changes of threshold in the sensory system, in systems affecting motivation, in the motor system, or in any combination of these. Threshold changes could be produced by genes acting on the nervous system itself—as in the hyperkinetic mutants of *Drosophila* described on p. 151—or more indirectly, by genes affecting metabolic rate or hormone secretion which will in turn affect the nervous system. In no case do we have proof, but sometimes the circumstantial evidence for direct action on the nervous system is strong. Thus the differences in the performance frequency of various sexual patterns between the inbred lines of guinea pigs mentioned earlier, remain even after injections of sex hormones[89] or thyroid hormones, which increase the basic rate of metabolism.[214] It seems most probable that the lines differ in the threshold properties of the sexual mechanisms in the brain.

It is not difficult to explain changes in the qualitative form or speed of patterns in terms of threshold changes also. Each fixed action pattern must have a co-ordinating 'centre' of some kind whose structure and properties are inherited and which, when activated, produces a stereotyped pattern of output to lower centres controlling muscles and groups of muscles. The centre calls into play each muscle group in the correct order, at the correct time and for the correct duration. This result must depend on a subtle series of threshold relationships both within the co-ordinating centre itself and between the various motor centres which control the muscles. The output of the co-ordinating centre may be completely 'pre-set', or it may be modified by feed-back from the muscles as the fixed action pattern is actually being performed. In insects we know centres, probably small groups of neurons, which control singing in crickets[120] and flight in locusts.[283] These produce quite normal output when isolated, although they can be modified by feed-back. Hoyle[119] discusses the different types of control in more detail in relation to his own work on insects, which have proved excellent material for such studies.

Genetic changes affecting thresholds either within the co-ordinating centre or one of the subordinate muscle groups will change the form of a fixed action pattern. For instance, muscles might be brought into action sooner or later in a sequence; they might be held on for a longer or a shorter time, or they might change the intensity and speed of their response as more or less of their constituent fibres are activated. Figure 7.5 shows a beautiful

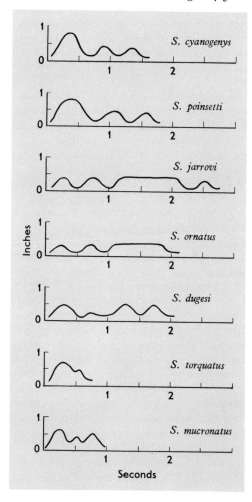

Fig. 7.5 The specific head-bobbing movements of some *Sceloporus* lizards. The movements of the head are represented as a line with height on the vertical axis and time on the horizontal axis. Changes to amplitude, speed and length of movements are all clearly shown. (From Hunsaker,[122] 1962, *Evolution*, **16**, 62.)

example of the behavioural results of such changes from Hunsaker's[122] work on lizards. Males of the genus *Sceloporus* show rhythmic head bobbing movements during courtship and also whenever they meet other lizards as a kind of species-identity signal. Each species has a characteristic pattern of bobbing which results from a series of contractions in the muscles which extend the front legs and thereby raise and lower the head and shoulders. Microevolution has produced several distinct versions by a series of changes of the type outlined above. This example is particularly clear because only two main groups of muscles are involved, but it is not difficult to envisage the differences in emphasis, illustrated in Fig. 7.4, evolving in just the same way.

We may conclude that genes affecting thresholds within the nervous system have been most important in the evolution of behaviour. Most of the amazing diversity of fixed action patterns we observe will have evolved by the accumulation of small quantitative changes, in just the same way as the body form of animals has evolved.

RITUALIZATION

Many of the fixed action patterns whose evolution we have been considering function as social signals and form part of threat, courtship or appeasement displays. We have already discussed the origins of such patterns in Chapter 5; they do not usually evolve *de novo*, but develop from 'intention movements'—especially of attack and escape—or displacement activities, of which breathing movements, preening, drinking or eating are amongst the most common. Such displays are called 'derived activities' by Tinbergen[259] to signify their origin from other types of behaviour, and the evolutionary process whereby they become modified to form social signals is called 'ritualization'. This term we owe to Julian Huxley[125] who first introduced it to describe the evolution of the elaborate mutual displays of male and female great-crested grebes, which we have already discussed (see p. 121 and Fig. 5.8). The display illustrated there has obviously been derived from elements of the grebes' nest-building behaviour, but some courtship displays have been so modified during the course of evolution that without comparative evidence from less modified species we would be hard put to suggest their origin. This is certainly the case with some of the movements used in the courtship of the tropical grass finches (Erythrurae) illustrated in Fig. 7.6. Male zebra finches (a) often perform beak-wiping on the perch during courtship—probably this is an un-ritualized displacement activity. In the related striated finch (b) and spice finch (c) the male performs a bow and holds himself with head lowered close to the perch beside the female. The similarity of this posture to a phase of beak-wiping is quite striking and Morris[196] suggests that the bow is a highly-ritualized pattern derived from the former.

Fig. 7.6 Beak-wiping in the courtship of three grass finches : (a) the zebra finch, in which the movement is unritualized and we see here the male just about to wipe his bill across the perch ; (b) the striated finch and (c) the spice finch, in both of which what is very probably the same movement has become ritualized. In these two species the male remains stationary in this position for some seconds and ritualized beak-wiping now looks rather like a bow. In all three cases the bird on the left is a female (Redrawn from photographs in Morris,[196] 1958, *Proc. zool. Soc. Lond*, **131**, 389.)

Ritualization involves all the threshold and frequency changes we have just been discussing and, in particular, changes in form and emphasis. Social signals must be conspicuous and most are extremely exaggerated in form and are accompanied by specially evolved releasers which enhance the effect of the movement (see the examples in Figs. 5.3 and 5.9). Good reviews of the ritualization of intention movements are provided by Daanje[60] and Andrew[9]; Tinbergen[259] and Morris[194] include other types of derived activities.

There are only two points connected with ritualization that require further mention here. Firstly, the concept of *typical intensity*, which term we owe to Morris.[195] If they are to be maximally effective, social signals must be clear-cut and unambiguous. For this reason many of them do not vary the form of the pattern to varying strengths of stimulus. A male cutthroat finch has a characteristic 'fluffed' courtship display posture. With weak stimuli he holds this posture only briefly but with strong stimuli he persists and begins his courtship dance. In either case, as shown in Plate 2, his display posture is almost identical, i.e. it shows a typical intensity. This situation may be compared with the threat posture of the cat (Fig. 5.6, p. 109) where the intensity of the response matches that of the stimulus. Postures or movements which have a typical intensity are more easily recognized but correspondingly convey less information about the sig-naller's motivational state. Both types of organization must have advant-ages and which of the two evolves presumably depends on details of the context in which the display has to function.

The second concept we must mention is the *emancipation* of ritualized patterns. The courtship of ducks, for example, includes patterns which are clearly derived from displacement preening (Fig. 5.9) and displacement drinking. Following the arguments developed in Chapter 5 (p. 115) it is possible that these patterns originally occurred by disinhibition as a direct result of a conflict between sexual and attack or escape tendencies when males courted females. One supposes that in ancestral males the appearance of preening or drinking depended on the conflict being suffici-ently balanced. Those males which performed these displacement activities most regularly were more successful in mating. In some way the appearance of these patterns served to arrest the female's attention and perhaps stimulate her sexually. However, in their descendants conflict may no longer be a prerequisite for the performance of preening and drinking patterns in the sexual situation. They are now almost invariable parts of courtship and highly modified in form. Tinbergen[259] suggests that they have become 'emancipated' from their original controlling mechanisms and are now controlled by the sexual mechanisms alone.

The two chief criteria for judging that a pattern is emancipated are (a) that it occurs out of the context which is appropriate for the presumed

(A)

(B)

Plate 2 The ruffled courtship posture of the male cutthroat finch. These two photographs illustrate the very small difference between the low intensity form (A), and high intensity form (B). The posture has developed a 'typical intensity' and, if shown at all, is shown nearly at its maximum, The small extent of the intensity differences can best be judged by comparing the two displaying males with the posture of the non-courting male on the lower left. The right-hand birds in both photographs are females (see p. 164). (From Morris,[195] 1957, *Behaviour*, **11**, 1.)

ancestral pattern, and (b) that fluctuations in the tendency to perform the presumed ancestral pattern have no effect on the tendency to perform the emancipated pattern. The first criterion is the same as that applied to justify any behaviour pattern being called a displacement activity, whilst the second predicts that, for example, the state of thirst should not influence the amount of display drinking a drake performs in courtship. We have little firm evidence that these criteria are met, but even so the reality of emancipation is in little doubt. Comparative evidence shows clearly the derivation of display drinking, but the pattern is greatly modified—in some species it is little more than a bob of the head—and it obviously has nothing to do functionally with water intake. At present we have little idea of what the emancipation process involves.

On p. 118 in our discussion of displacement activities, we mentioned the need for a relatively stable balance between conflicting tendencies if a displacement activity was to have time to 'break through'. Such a balance can never stay perfect for long and presumably there is a range of conflict states in which equality of the two tendencies is sufficiently close to allow another behavioural system to manifest itself briefly—Rowell[224] refers to this range as 'effective equality'. One initial step in ritualization might be to widen the range of effective equality between two conflicting tendencies and thereby increase the frequency of a displacement activity. Blest[32] discusses in more detail some problems of emancipation and the ritualization concept generally.

So far in this discussion of ritualization we have applied the term solely to the evolution of courtship and threat displays, for which it was used originally. It is worth noting, however, that the opportunistic nature of selection whereby an element of behaviour occuring in one situation serves as raw material for the ritualization of another pattern, is also seen in other contexts.

Many tropical fresh-water fish, such as gouramis and fighting-fish, build nests of air bubbles. A male takes a mouthful of air at the surface and blows bubbles coated in saliva, which he builds up into a cluster clinging to the surface film amongst water weed. He deposits the fertilized eggs into the bubble nest after courtship and mating. Braddock and Braddock[35] suggest that the construction of bubble nests has followed on the evolution of air breathing. This habit evolved in quite another connection, as a response to the low oxygen content of water in tropical rivers and lakes, but it meant that the fish was constantly taking mouthfuls of air at the surface. Usually when it does so a few bubbles escape from its mouth and remain intact at the surface film for a time. From this point onwards it is not difficult to imagine the stages whereby bubble nests became well-oxygenated shelters for the developing eggs.

Another remarkable example of evolutionary opportunism is the 'dance' of the honey-bee, first analyzed by von Frisch.[77] By means of its pattern of movement on the comb, a returning forager can indicate to other worker bees the direction, distance and quality of the food source upon which it has been feeding. Without going into the details of von Frisch's classic experiments, the bee transposes the angle between the sun (which it has been using as a compass) and the flight path from hive to food source, to the same angle with relation to gravity on the vertical surface of the comb. In the convention adopted by bees during the course of evolution, directly towards the sun transposes to directly upwards on the vertical comb. The dance takes place in almost total darkness and other bees perceive the angle to gravity by following closely the movements of the dancing bee and thus making their own path with respect to gravity correspond to hers. When they leave the hive they make the same transposition in the opposite direction, angle to the vertical becomes angle to the sun.

We do not know the exact manner in which the dance 'language' evolved, nor do we have living representatives of any intermediate stages that help much (see Lindauer[153A]). However one thing is certain; there is nothing exceptional in principle about the transposing of orientation with respect to the sun into orientation, at an equivalent angle, with respect to gravity. These two physical features are the most dominant in orientation and other insects which have been investigated also transpose between them. Thus ants which have been running at a particular angle with re- lation to a light on a horizontal board, change course abruptly to take up a related angle with respect to gravity, when suddenly the light is switched off and the board tilted into the vertical plane (see Vowles[269B]). Certain beetles behave in an analogous way. It is probable that honey bees have modified an orientation system which is common to all insects. So far as we know no other insects use this ability for communication, but the potential was there. Given the demands for maximum efficiency in foraging by honey bee colonies, natural selection has been able to elaborate this remarkable means of communication from relatively mundane and in- conspicuous origins.

SEXUAL ISOLATION

Related species do not normally hybridize under natural conditions because it is disadvantageous for them to do so. Hybrids, although they may show some signs of 'hybrid vigour', are usually sterile or sub-fertile. Even if they are fertile they are rarely as successful as either parental type. The latter each have a set of genes which has been selected over many generations as the best for their own environment. The hybrids inherit a compromise set of genes which does not equip them so well in either parental environment. We know several cases in plants where hybrids

between two species have invaded new environments and been very successful but this is very rare, if it occurs at all, among animals.

Consequently there is a strong selective advantage to choosing a member
of one's own species for a mate. This is particularly true for females; males
can usually mate several times in their lifetime but some female insects, for
example, mate only once. If they mate with the wrong type of male they are
effectively sterilized. Accordingly it is common to find that discrimination
is stronger by females than by males. Females cannot afford to take risks
but males can, and indeed too much discrimination may be actually disadvantageous for them. It will usually be better for a male to risk the oc-
casional infertile mating than ever to miss the chance of a fertile one. In
Drosophila, for example, females are more selective than males and it is
often their active rejection of foreign suitors which prevents interspecific
hybridization.

Sexual isolation may be defined as 'behavioural barriers to hybridization
between species or populations'. It forms one aspect of the more general
phenomenon of reproductive isolation, which is fully discussed by Mayr,[183]
and is one of the most important ways in which behaviour affects the
evolution of animal populations.

The degree to which two related species will meet under natural conditions varies greatly. Even if they co-exist in the same general area, two
species will rarely live in exactly the same part of the habitat; competition
is bound to drive them into specialization. In Britain, chiff-chaffs (*Phyllo-
scopus collybita*) and willow-warblers (*P. trochilus*) inhabit the same woods
and both nest on the ground. However, for feeding the chiff-chaff moves
amongst the high trees and the willow-warbler in the lower branches of
trees and bushes. Clearly they are specializing, and in the Canary Islands
where the willow-warbler is absent, the chiff-chaff moves into the willow-
warbler's niche as well.

Obviously the selection of a habitat characteristic for the species will
form part of normal reproductive isolation. *Drosophila pseudoobscura* and
D. persimilis are two closely related species whose ranges overlap extensively in the south western part of the United States. They can be
hybridized in the laboratory, particularly at low temperatures but although
many tens of thousands of flies have been caught in the wild and examined,
no natural hybrid has ever been found. The two species seek out different
micro-habitats within an area, *pseudoobscura* drier and lighter places,
persimilis cooler and moister, and their preferences keep them effectively
isolated. Even in the laboratory, *pseudoobscura* females will not readily
accept *persimilis* males or *vice versa* because their courtship songs are so
different, (see p. 153). In most animals isolation by habitat is supported
by the evolution of highly specific signals which enable animals to detect
their own species, sometimes at a considerable distance.

In the crickets, grasshoppers and cicadas among insects, and the frogs and toads among vertebrates, the males have 'assembly calls' which attract females in the mating season (see Alexander[1] for insects and Blair[29] for Anura). A number of different species may live in the same area, and with frogs, males of several species may be calling from the same small pond or ditch. It is remarkable how distinctive is each species' call and females respond only to the call of their own males. There can be little doubt that the diversification of calls is a direct result of selection in many cases. For example, Blair[28] describes how two species of frog, *Microhyla olivacea* and *M. carolinensis*, have a wide range in the southern United States, the former more to the west and the latter more to the east. At the extremes of their ranges where only one species is present their assembly calls are quite similar. More centrally where both species overlap and use the same ponds for breeding the calls have diverged and are quite distinctive. Hybrids are occasionally found in nature and Blair has shown that, apart from calls, there is little other barrier to hybridization. The males, like those of all Anura, will clasp and attempt to mate with almost any object of the right size and the females, if placed with males of the wrong species, will accept them readily.

Perdeck[206] describes a similar example in the two closely related grasshoppers, *Chorthippus brunneus* and *C. biguttulus*. These are very similar morphologically and occur in the same areas but their songs are very distinctive and, once again, females are attracted only by the song of their own males. If they are lured to foreign males by tape recordings of their own species, courtship and copulation proceed normally and hybrids are produced.

In cases such as these just quoted it seems likely that two populations with similar assembly calls came to overlap after many generations of geographical isolation, during which time they had diverged in a number of other ways. Selection then favoured the divergence of their calls because hybrids were at a disadvantage. Alexander and Bigelow,[3A] describing the calling songs of the large variety of crickets and grasshoppers which live together in the eastern United States, find that—with one exception—they all have different calling songs. (Sometimes species are so similar morphologically they have been first distinguished by song.) The exception is the pair of cricket species *Acheta pennsylvanicus* and *A. veletis* which live in exactly the same area but have identical calling songs. However this is the exception that proves the rule, because *veletis* and *pennsylvanicus* mature in spring and autumn respectively, and therefore males are calling and attracting females to mate at completely different times of the year. These two species have diverged from a common ancestor by specializing in the early or late timing of their breeding cycle. The fact that during this evolution, their songs have remained identical indicates that unless there is active selection for divergence calls will remain very constant, presumably because variants are likely to be 'misunderstood'. The most compelling evidence for this conclusion is provided by three crickets of the genus *Gryllus*, *G. campestris* from Europe, *G. bermudiensis* from Bermuda and *G. firmus* from the eastern United States. They must have been isolated from each other for many thousands of generations and they have diverged considerably in structure, but their calling songs remain almost identical.[3]

There is mutual adaptation between a male's call and a female's response. If selection favours the divergence of calls then changes to the behaviour of one sex will impose selection on the other to respond accordingly. Sudden, large changes will tend to be disadvantageous because they will be too 'difficult' for the opposite sex to follow.

Assembly calls are only one type of sexual isolation mechanism. Other groups of animals discriminate their own kind by scent, colour pattern or some combination of features. Colour pattern is obviously important in some birds; two or three species of ducks may gather and court on the same pond but the males each have highly distinctive patterns.

Scent differences are of great importance for sexual isolation in insects. Some virgin female moths 'assemble' males to their highly specific scents.[233] In *Drosophila* stimuli received when a male approaches and taps the female with his fore-tarsi provide chemical evidence of identity for both sexes.[176]

Courtship behaviour itself, if we except song patterns, is probably used less often for identification. The differences between close relatives are, as we have seen, mostly quantitative. It seems that selection for divergence can operate most efficiently to change colour or scent in a clear-cut distinctive way. Courtship patterns may help in sexual isolation in that the movements and postures used may serve to display specific colours or scents in a conspicuous fashion.

With sexual isolation we are made most aware of the evolution of differences in behaviour for their own sake as an aid to discrimination. It is worth noting, however, that sometimes selection favours the opposite tendency and will cause animals to evolve similarities, again for their own sake, with other species. Where colour patterns are concerned, the phenomenon of Müllerian mimicry, is a familiar example. Many distasteful insects (e.g. caterpillars of the cinnibar moth, ladybird beetles) or those with stings (e.g. wasps, bumble-bees) are coloured with conspicuous stripes or blotches of black and yellow or black and orange. The same colour patterns are found in some fish, amphibians, lizards and snakes—in every case associated with powerful defence mechanisms of some kind. Selection will favour a distasteful or poisonous animal having a conspicuous colour and also showing conspicuous behaviour, or at least not trying to hide itself. Predators learn rapidly as a result of one unpleasant experience and avoid contact with others of the same type. Müller suggested that it was no coincidence that so many diverse animals used the same type of warning colouration. The obvious conclusion is that they have evolved to resemble one another in order to benefit from each other's warning effect. A predator avoiding one type of distasteful prey will also avoid others without having to sample and learn each one separately—clearly an advantageous situation for the prey species.

Convergence of colour markings is also found among bird species which form mixed flocks for feeding (we shall discuss the significance of flocks more fully in Chapter 10). Moynihan[196D] describes a number of examples from the American tropics, e.g. tanagers, warblers and finches of western Panama which form mixed flocks and are largely black and/or yellowish sometimes with white markings. Birds from these species tend to assemble and move together and the common markings are probably specially evolved to maintain the cohesiveness of the group. Moynihan refers to this as 'social mimicry'.

In Chapter 3 we mentioned that many small birds respond to the alarm calls of other species as well as their own, and that these calls have converged during evolution to provide a common sign stimulus. Marler[179] discusses the characteristics of an 'ideal' alarm call which must carry as far as possible whilst providing the least clues to a predator on the localization of the calling bird. Amongst other things this means that the call should be of constant pitch and both begin and end gradually. It is the abrupt onset of sounds which allows their localization, since one way vertebrates locate a sound source is by comparing the time at which the sound hits each ear. Fig. 7.7 shows the sound spectrograms of alarm calls in small birds from three different families which are extremely similar and possess the properties outlined above. They provide a striking contrast to Fig. 7.8 in which we see the territorial display songs of three species of European warbler. Here the requirements of advertisement and sexual isolation have led to marked divergence between close relatives living in the same area.

EVOLUTIONARY COMPROMISE

In conclusion it is worth stressing that natural selection is opportunistic and operates upon whole animals not upon parts. With behaviour, as with any other character, selection makes use of the variation that is available and produces, not the maximum possible response from the character, but the best compromise with all the others.

We noted in Chapter 3, p. 47, that it was often possible to make an artificial releaser which was more effective than the natural one. The latter may be part of a butterfly's wing, for example, which also has to serve for flight and selection has to compromise at a level lower than maximum. We can see a similar situation with the nuptial colours and songs of male birds. The bright plumage and loud song which attract a mate are also likely to attract a predator. Most males stop singing as soon as their mate begins to incubate and some lose their breeding plumage at the autumn moult. There must obviously be a compromise between reproductive success and danger from predation. This will be struck at different levels depending on other circumstances. Polygamous species like the pheasants, where not all males are successful and one may mate with several females,

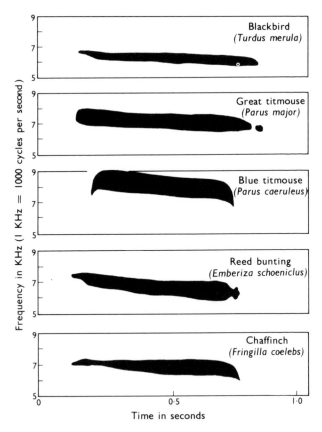

Fig. 7.7   Sound spectrographs of the alarm calls of five species of passerine bird from three different families. They are remarkably similar in form and pitch and they share the property of being difficult to locate. (From Marler and Hamilton,[179B] 1966 *Mechanisms of Animal Behavior*. John Wiley, New York and London.)

have developed very extravagant reproductive plumage and keep it all the year round. However, male pheasants have nothing to do with incubation or rearing young which is the task of the cryptically coloured females. A brilliantly coloured male who attracts and mates with six females in one season before falling prey to a fox, will leave more descendants than a more sober male who lives for 10 years but never attracts a mate!

Polygamy is only possible in animals where the female is capable of protecting and feeding the young by herself. This is most often the case with herbivores because their food is relatively abundant. Thus polygamy is the rule amongst the whole order of herbivorous game birds—the

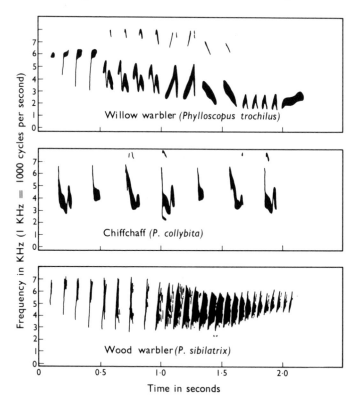

Fig. 7.8   Sound spectrographs of the songs of three very closely related species of European warbler. The three are very similar in plumage and share the same habitat, hence the songs probably serve, in part, as a means of sexual isolation. They are strikingly different in form and quality but share the property of being easy to locate (From Marler and Hamilton,[179B] 1966 *Mechanisms of Animal Behavior.* John Wiley, New York and London.)

pheasants being one example. It is also common amongst herbivorous mammals such as deer. Amongst insectivorous or carnivorous animals the male is usually needed to help gather food for the young, consequently monogamy is the rule and the males rarely show extravagant plumage or structures equivalent to the antlers of the polygamous deer. The males of monogamous species have to strike a more functional balance between advertisement and parental requirements.

The work of Tinbergen and his group on gulls has revealed several beautiful examples of the way in which their behaviour has been forced to change to fit in with other environmental circumstances. For example, Cullen's[57] study of the kittiwake (*Rissa tridactyla*) has revealed numerous differences from the other gulls. The kittiwake has taken to nesting on narrow cliff ledges and by this means protects its eggs and young from predators which take a considerable toll from the ground-nesting gulls. However, suitable ledges are in short supply and there is intense competition for nest sites. Prolonged fights involving both males and females take place on the ledges and, beaks locked together, kittiwakes often fall into the water where they continue to beat one another with their wings! Fights of this severity and persistence are virtually unknown in the other gulls. The high level of aggressiveness in kittiwakes affects not only their fighting but also their courtship displays. These show a high proportion of aggressive elements and overt aggression between male and female persists late into the breeding season. To counteract this aggressiveness, appeasement displays are also well developed (see p. 111). Here they consist of turning the head away from the opponent and depressing the beak—an example of the 'anti-threat' type of appeasement we discussed in Chapter 5. The commonest form of threat is 'choking', also described in Chapter 5, where we considered evidence that when this posture is adopted by other gulls—in which it is far less common—it indicates that the bird is strongly motivated to stay put. The kittiwake's attachment to and defence of its nesting ledge has to be most intense. Accordingly selection has favoured a compromise between attack and escape tendencies which is more biassed towards attack than in the other gulls.

Cullen's study revealed a number of other behavioural modifications both in adult and young which could all be traced back to the original shift to cliff nesting. More recently Cullen and Ashmole,[59] and Hailman[93] have shown a similar series of adaptions in a cliff-nesting tern, the black noddy, and a second cliff-nesting gull, the Galapagos swallow-tailed gull, respectively.

8

Learning—1

Throughout this book we have had to make reference to learning as it affected the particular aspect of behaviour under discussion; now attention must be focused on learning itself. In this chapter we shall consider the general phenomenon and the different forms it takes, together with some account of how learning ability varies between different animal groups. The concluding chapter will concentrate on the mechanisms of learning.

Thorpe[255] defines learning as '. . . that process which manifests itself by adaptive changes in individual behaviour as a result of experience'. This definition draws attention to two important features. Firstly, learning normally results in *adaptive* changes, and in Chapter 2 we discussed how learning and instinct are both ways for equipping an animal with a set of adaptive responses to its environment. Normally both methods are found in combination and logically they have much in common. In one case we have the selection of individuals bearing the best genes during the evolution of a population of animals; in the other, the selection of the best responses during the learning process in an individual animal. Russell[226-9] develops the analogy between learning and evolution in a most complete and revealing fashion.

The second important point arising from Thorpe's definition is that, strictly speaking, learning is a *process* which we cannot usually measure directly; we measure what has been remembered as a result of learning. Because we can communicate so easily with human subjects, they are, in many respects, better material for learning studies than animals. For instance, it is possible to test our memory in two ways; by 'recall', i.e. by reciting or writing down a list of nonsense syllables which we have previously learnt, or by 'recognition', i.e. presented with a set of nonsense syllables which includes those we have learnt, we record which syllables we recognize. Recognition is always an easier task than recall, because the situation provides stimuli which, as we say, 'jog our memory' and helps the process of recall. If we have trained a rat to run through a maze we cannot ask it to draw a map of its route on a piece of paper. The only way to test what it has learnt and retained is to put it back in the maze and observe its behaviour. If the rat makes mistakes we have no means of knowing whether it failed to learn adequately or learnt, but failed to recall.

We shall return to the problem of separating learning from recall in the next chapter. Here we must accept the inevitable limitations of measurement and concentrate on describing the conditions under which animals learn and what they can learn.

Animal learning studies have, until recently, been dominated by the work of experimental psychologists. Soon after J. B. Watson founded the new 'behaviourism', the learning abilities of the domestic white rat became the subject of a scrutiny which has continued for half a century. Out of this work have grown a number of schools of learning such as those associated with Hull, Tolman and Skinner. Each school has tried to construct a system of behavioural laws which will predict more or less exactly under what conditions learning will occur. We shall not discuss their respective merits at any length; there are admirable accounts of learning theories in Munn's[197] classic textbook on the rat, and in Broadbent.[36] Until quite recently one weakness they have shared has been excessive concentration upon one or two domesticated species—notably the white rat and the pigeon.

CLASSIFYING LEARNING

Learning occurs in a great range of different animals under a wide variety of circumstances. Central to some schools of thought in psychology has been the idea that there are general 'laws of learning' which apply with equal force wherever it occurs. This viewpoint has taken some hard knocks recently (see Seligman's[236B] review and the discussion beginning on p. 185) but even if it were true it is none the less helpful to try to distinguish different categories of learning. This is a convenient way to organize our thinking, although we must certainly recognize that any classification is likely to be little more than a set of artificial abstractions. They may not correspond to the natural situation, particularly if a broad range of animal types is under consideration.

The classification we shall use here is taken from Thorpe[255] and the account which follows owes a great deal to his book, which includes much more detail and full references. The particular value of Thorpe's book lies in his complete coverage of the whole animal kingdom. His classification of learning, together with the commoner synonyms found in the literature, is

1. Habituation

'Associative learning'
{
2. Conditioned Reflex (CR) Type I (='classical conditioning' or 'respondent conditioning')
3. 'Trial and error' and CR Type II (much of which is called 'instrumental conditioning' or, by Skinner, 'operant conditioning')
}

4. Latent learning
5. Insight learning
6. Imprinting

Habituation

It is convenient to deal with habituation first because, in some respects, it is the simplest form of learning. Unlike the other forms, habituation involves not the acquisition of new responses but the loss of old ones. If an animal is repeatedly given a stimulus which is not associated with any reward or punishment, it ceases to respond. Birds soon ignore the scarecrow which put them to flight when it was first placed in a field. A snail crawling across a sheet of glass retracts into its shell when the glass is tapped. After a pause it emerges and continues moving; a second tap causes retraction again but it emerges more quickly. So we proceed with the snail's response becoming increasingly perfunctory until it ceases to respond at all.

A study by Clark[51A,51B] on the ragworm, *Nereis*, illustrates some of the typical features of habituation. *Nereis* is a marine worm which normally lives in a burrow or tube which it constructs in the mud at the bottom of brackish estuaries. The worm's head and anterior segments protrude from the tube whilst it feeds from the surface of the mud. At such times a variety of sudden stimuli will cause the worm to jerk back rapidly into its tube. In the laboratory Clark could easily get the worms to live in glass tubes in shallow basins of water. He found that jarring the basin (mechanical shock), touching the head of the worm, a sudden shadow passing over and a variety of other stimuli would all cause rapid retraction into the tube, but the majority of worms emerged again within a minute. If these stimuli were repeated at 1 minute intervals the proportion of worms responding fell off until none of them were retracting. Clark found that habituation occurs more rapidly if stimuli were given close together. For example, with a bright flash of light it took less than 40 trials at half-minute intervals, but nearly 80 trials if the interval was 5 minutes. The speed of habituation also depended on the nature of the stimulus; mechanical shock, shadow, touch and light flash all produced their characteristic rates of habituation. Further habituation was to a large extent 'stimulus-specific'; Fig. 8.1 shows how the waning of retraction to repeated mechanical shock is independent from that to a moving shadow. In this, the behaviour of the worms resembles that of the nestling chaffinches in Prechtl's[207] experiments described on p. 10.

There are a number of other processes which may be confused with habituation because they also lead to a reduction in responsiveness. Results such as those illustrated in Fig. 8.1 eliminate any possibility that the waning of response is due to motivational changes or to muscular

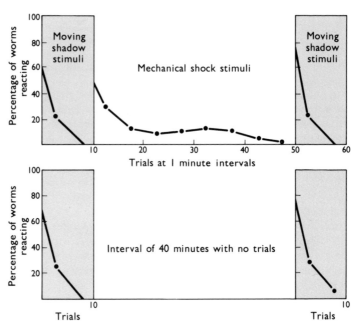

Fig. 8.1 The rates of habituation to two different stimuli in *Nereis*. The response measured is the sudden retraction of the worm into its tube and trials are given at 1-minute intervals. The shaded areas record the responses of a group of 20 worms to a moving shadow. Within 10 trials they have all ceased to respond, but switching to a mechanical shock stimulus (unshaded areas in upper graph) brings back the response in half the worms, and it characteristically habituates more slowly, taking more than 30 trials. The recovery of the response to moving shadow is complete after 40 minutes, whether the habituation trials to mechanical shock intervened (upper graph), or if the worms were simply left alone (lower graph). Clearly habituation is quite independent for these two very distinct types of stimulus. (From Clark,[51A] 1960. *Anim. Behav.*, **8**, 82.)

fatigue, but it is often more difficult to eliminate sensory adaptation. Many sense organs eventually stop responding to repeated stimulation—on p. 65 we discussed the adaptation of the chemo-sensory hairs in the blowfly. *Phormia*. We cease to be aware of our clothes within a minute or so of putting them on, because the tactile receptors in the skin cease to respond. In *Nereis*, Clark could also eliminate sensory adaptation as an explanation for the waning. For example, the worm soon ceased to retract when touched by a probe, but clearly still detected the stimulus because it then attempted to seize the probe with its jaws.

Sensory adaptation is usually a short-lived phenomenon; a few minutes without stimulation is usually sufficient for complete recovery. It would not be appropriate to say that we habituate to the feeling of our clothes, and we retain the term for a more persistent waning of responsiveness which must be a property of the central nervous system and not the sense organs.

Clark could detect some recovery from the waning of retraction within an hour or less in *Nereis*, and the worm was completely recovered within 24 hours. We have already mentioned that this waning was stimulus-specific, and it is this type that meets most of the generally accepted criteria for habituation. However there are other types of waning with different time courses and it is difficult to make any firm distinctions on behavioural grounds. Hinde's experiments on the mobbing behaviour of chaffinches described in Chapter 1 (see Fig. 1.5 for example) suggest that there is a dual recovery process following waning. There is a relatively rapid component, complete within 24 hours and also stimulus specific and thus very comparable to the *Nereis* situation. But there is also a long-term component which showed little sign of recovery over the same period, because the response remained greatly reduced. It is doubtful if we are justified in lumping these two components under the term 'habituation'. In nature sensory adaptation, short and long term waning will probably all occur together in some situations, so it is impossible to define habituation with any degree of rigidity.

Habituation-like phenomena are found in every group of animals from the Protozoa upwards. They appear to be almost a general property of living material and certainly all the typical characteristics of habituation and recovery can be exhibited at the level of single neurons and neuro-muscular junctions (see Bruner and Kennedy,[39A] and reviews in Horn and Hinde[118A]).

Habituation is important process for adjusting an animal's behaviour to its environment. Small prey animals such as *Nereis* cannot spend too long sulking in their tubes—they must normally be out feeding. Although it is adaptive to retreat rapidly to sudden shadows, not every shadow means a predatory fish is overhead, it might just as well be a floating piece of sea weed. In general, frequent shadows are more likely to be sea weed than predators and it is adaptive to cease responding when a repeated stimulus has no attendant consequences.

Habituation also plays an important part in the development of behaviour in young animals which are often threatened by a wide range of predators and may begin by showing escape responses to anything large which moves. Rapidly they learn to ignore leaves moving in the wind and other neutral stimuli. Schleidt's[232] work on the alarm response of young turkeys, described on p. 57, illustrates this point well. The inherited pecking response of newly-hatched chicks is at first directed towards any small object that contrasts with the background. Again responses to inappropriate objects quickly habituate and at the same time the chick learns positively which objects are food.

Conditioned reflex type I

The term 'conditioned reflex' is inseparable from the name of the great Russian physiologist, I. P. Pavlov,[204] whose school was active around the turn of the century. Pavlov's influence on Russian behaviour studies is still very great but, perhaps because the behaviour theory he developed has attracted little favour here, his influence in the West has been less. Pavlov and Sherrington were working at the same time but from completely different viewpoints. Sherrington studied the organization of reflexes in the isolated spinal cord of dogs and cats, having deliberately cut off influences from the higher centres. Pavlov worked with intact animals and considered that, just as simple reflexes are a property of the spinal cord, so conditioned reflexes are the particular property of the higher centres of the brain, especially the cerebral hemispheres. Pavlov's aim was to study 'the physiology of higher nervous activity', but most of his experiments were, in modern terms, pure experimental psychology. Indeed, Pavlov was really one of the founders of experimental psychology; he was applying objective techniques to the study of learning some years before J. B. Watson.

Pavlov's classical experiments with dogs often involved the 'salivary reflex'. Dogs salivate when food is put into their mouths and Pavlov could measure the strength of their response by arranging a fistula through the cheek from the salivary duct, so that drops of saliva fell from a funnel and could be counted. A hungry dog was placed on a stand, restrained by a harness and every precaution was taken to exclude disturbances. In this position it could be given various controlled stimuli such as lights, sounds or touch, and meat powder could be puffed into its mouth through a tube. A standard quantity of meat powder caused the secretion of a certain amount of saliva. Now Pavlov preceded each ration of powder by, say, the sound of a metronome ticking. At first, this stimulus caused no response, save that the dog pricked up its ears momentarily. However, after five or six pairings of metronome followed by food, saliva began to drip from the dog's fistula soon after the metronome started and *before* the meat powder arrived. Eventually the amount of saliva produced to the metronome alone was the same as that which was given to the meat powder.

The dog had learnt to respond to a new stimulus which was previously neutral, and Pavlov called this the 'conditioned stimulus' (CS). The salivation response to the CS is the 'conditioned response' (CR). Prior to learning, only the meat powder or 'unconditioned stimulus' (UCS) produced salivation as an 'unconditioned response' (UCR).

Pavlov found that almost any stimulus could act as a CS provided that it did not produce too strong a response of its own. With very hungry dogs even painful stimuli, which initially caused flinching and distress, quite soon evoked salivation if paired with food.

The CR is formed by the *association* of a new stimulus with a reward or 'positive reinforcement' (a term which we shall examine more closely later). A CR for withdrawal can also be formed by associating the CS with a punishment or 'negative reinforcement'. An electric shock to the foot causes a dog to lift its paw; if a metronome is paired with the shock, the dog soon raises its paw to the sound alone.

Conditioned reflexes of this type have been observed in many different animals from arthropods to chimpanzees. For example, birds learn to avoid the black and orange caterpillars of the cinnibar moth after one or two trials which reveal their evil taste. They associate this with the colour pattern and generalize (see p. 186) from cinnibar caterpillars to wasps and other black and orange patterned insects. Because predators generalize it is advantageous for different distasteful insects to resemble one another—the phenomenon of Müllerian mimicry (see p. 170).

In nature we rarely observe a conditioned reflex in as 'pure' a form as in the laboratory. Bees do not just learn to associate a colour with the nectar reward, they also learn the position of the group of flowers with respect to their hive and learn at what time of day the nectar secretion is highest. Even Pavlov in his scrupulously controlled environment found that his dogs learnt more than one particular response to one particular stimulus. A hungry dog familiar with his laboratory's methods would run ahead of the experimenter into the test room and jump up on to the stand with every sign of expectancy.

Trial and error

In the CR type 1 the animal starts out with a *response*—the UCR linked to its UCS—which subsequently becomes attached to a novel stimulus; hence the term 'respondent conditioning' which is sometimes used for this type of learning. In contrast to this situation we may have one in which the animal is motivated by thirst, hunger or fear as before, but has no UCS to evoke an appropriate UCR. Rather, the animal is exploring or showing appetitive behaviour, during the course of which it performs spontaneously a variety of motor patterns, sniffing, walking, looking round etc. Suppose now one of these patterns is followed by reinforcement—e.g. a hungry animal receives food—then, if this association is repeated a number of times, the animal learns to perform this pattern regularly in this particular situation.

An example will make this clearer and suggest why this type of learning is justifiably called 'trial and error'. In his pioneer experiments on learning Thorndike used a variety of 'problem boxes'; one of these is shown in Fig. 8.2. It is a cage which can be opened from inside only by depressing the lever. A cat is shut in and tries hard to escape, it moves around rest-

Fig. 8.2 One of Thorndike's problem boxes. A cat is confined inside the cage and must learn to pull the string loop to open the door. (From Maier and Schneirla,[173] 1935, *Principles of Animal Psychology.* © 1935. By permission of McGraw-Hill, New York and Maidenhead.)

lessly and after a time—by chance—it steps on the lever and the door opens. The second trial may be a repetition of the first, and the third, but soon the cat concentrates more attention on the lever and eventually it moves swiftly across the box and presses the lever as soon as it is confined. The name 'trial and error' obviously fits this type of learning very well. The cat learns to eliminate behaviour which led to no reward and increases the frequency of behaviour which is rewarded, but in the early stages there is little system in its activity—the first reward is obtained by pure chance.

We have already described the use of a Skinner box, which is basically a problem box of a convenient form in which an animal learns by trial and error that pressing a bar yields a small reward. Because the animal's own 'spontaneously generated' behaviour has been instrumental in its gaining a reward, such learning is often called 'instrumental conditioning' (Skinner[244] also uses the term 'operant conditioning'), but it is no different in principle from trial and error. In this type of conditioning there is no CS, unless we consider that the proprioceptive feedback from the muscles during the performance of the CR—bar pressing for instance—is recorded as a pattern of stimulation which signals the reward. In the course of time other features of the trial and error situation may come to act as conditioned stimuli which signal to the animal that the response is now 'appropriate' and will yield reward.

It is not at all clear how fundamental is the distinction between trial and error and the CR Type I. Konorski[142] has argued that the former is best regarded as a CR Type II and also refers to this class a special type of learning in the Pavlovian situation. A dog on the stand has its left front foot lifted and then meat powder is blown into its mouth. After a few trials the dog *spontaneously* lifts its foot when hungry and placed on the stand. Here we are simply directing the nature of the animal's CR, and the problem box situation in which the animal spontaneously performs the rewarded act is obviously more natural.

Trial and error learning will often be involved when animals modify their appetitive behaviour to obtain food, shelter or a mate. Most commonly it will be mixed with classical conditioning (CR Type 1) because both new stimuli and new behaviour patterns must be learnt. Learning to run a maze is a familiar example when a rat may learn to use a combination of clues, some visual depending on stimuli perceived at the choice points, others 'kinaesthetic', i.e. learning whether to turn to right or left.

Trial and error is probably the most appropriate category under which to classify the learning of new motor skills. Young mammals and birds, for example, may perfect the co-ordination of their movements by practice, often during sessions of play with their siblings or parents.

Some characteristics of associative learning

Because both types of associative learning have much in common and are relatively simple, this is a good point at which to summarize what general features of the learning process they exemplify. Most of these features were first clearly described by Pavlov from his work on the CR Type I. As mentioned earlier there are real problems in drawing up 'laws of learning' which have general validity and this account will have to be highly simplified. Where there are marked deviations from the general rule they will be noted.

Contiguity

In the great majority of conventional learning situations it has been found that if a conditioned stimulus or a new response, such as lever press-ing, is to become associated with a reinforcement, the two must occur close together. Figure 8.3 shows some of Pavlov's findings. Conditioning occurs most readily when the CS starts before the UCS and goes on to overlap it. It is very difficult to get a CR if the CS ends more than about a second be-fore the UCS begins, or if the CS goes on after the UCS has stopped.

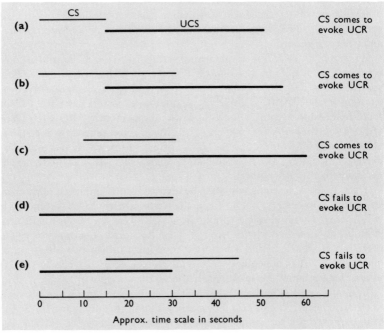

Fig. 8.3 The effect of the sequence of stimuli upon the formation of a condi-tioned reflex. In each case the upper, thin line denotes the duration of the condi-tioned stimulus (CS), the lower, thick line that of the unconditioned, reinforcing stimulus (UCS). The results are given on the right ; note that the CS must not end with or persist beyond the UCS if a positive conditioned reflex is to be established. If it does so, then the CS will remain neutral and may even tend to inhibit the response to the UCS. (From Konorski,[142] 1948, *Conditioned Reflexes and Neuron Organisation*. University Press, Cambridge.)

Now in trial and error or instrumental conditioning the CR always *precedes* the reinforcement but, again, the latter must follow rapidly if learning is to occur. Skinner[244] found that delays of only a few seconds between a rat pressing the bar and the delivery of food greatly slowed the rate of learning the response. The deleterious effects of delayed reward can often be overcome by introducing a 'secondary reinforcement'. Suppose the rat learns that a reward is delivered when a light comes on in the box (light and reward must overlap as discussed above), then it will learn to press the bar to switch the light on. Light becomes a secondary reinforcement or a 'bridging stimulus' between the response and the primary reinforcement—food. Bridging stimuli are useful for training animals—as in a circus—when it is often difficult to present a reward immediately after the response is made.

However there are certain types of associative learning which consistently occur when reinforcement is delayed for a matter of hours. Barnett[14] describes how wild rats only nibble at small amounts of any novel foods that appears in their territory. If it proves edible, they will gradually take more on successive nights until they are eating normally. If it is poisonous, and they survive, they avoid it completely on subsequent occasions. This type of behaviour is highly adaptive, and makes poisoning rats no straightforward task. The interesting feature for our discussion is the delay that must ensue between a rat tasting poison bait (always made superficially palatable with sweet substances) and any subsequent ill effects. Few rat poisons take less than an hour to produce effects. Laboratory findings have confirmed this ability, Not only will rats learn to avoid tastes associated with sickness that sets in at least an hour later, if deprived of the vitamin thiamine they will learn to choose a diet containing it, although many hours must elapse before they can feel its benefits.

It is only when a new *taste* acts as the CS that reinforcement can be so delayed. In one ingenious experiment, Garcia and Koelling[80A] supplied rats with a drinking tube containing saccharin-flavoured water so arranged that when they licked the tube, bright lights flashed on. During these sessions the animals were irradiated with X-rays which made them sick about an hour later. Subsequently the rats avoided saccharin taste but did *not* avoid flashing lights. Conversely if they were given flashing light plus immediate electric shock to the feet whenever they licked the tube, they subsequently avoided the light but still licked at saccharin.

Rats are in some way 'prepared' to associate taste with sickness after a single trial and a long delay, but visual stimuli and sickness are not connected. Conversely light and electric shock are easily associated if they occur close together, but taste and shock are not.

In the past it has commonly been assumed that contiguity plus reinforcement is all that is required to associate any stimulus with any response. However experiments such as these (and many more ably reviewed by Seligman[236B]) suggest that animals approach learning situations with a good deal of built-in bias. This bias usually relates to the natural requirements which their behaviour has evolved to meet. We discussed in Chapter 2 (p. 36) that animals may inherit a tendency to learn particular things. Breland and Breland[35A] in a paper which they mischievously entitled

'*The Misbehavior of Organisms*' (note Skinner's title) give a number of examples from their own efforts to train animals to perform tricks for commercial purposes. They found that the laws of contiguity and reinforcement are not sufficient to overcome powerful inborn tendencies to behave in particular ways. Thus chickens persisted in scratching at the ground even though it hampered their getting a food reward.

Repetition

Although, as we have seen, one-trial learning can occur especially with strong punishment as a reinforcement, most associative learning improves with repetition. Pavlov found the amount of saliva produced to the CS steadily increased with each trial which was reinforced until it was at the same level as that given to the UCS. The rat learning a maze steadily reduces its errors with successive trials until it runs unhesitatingly through to the goal box where food is found. The number of errors or the time taken to reach the goal box on each trial can be used to plot learning curves such as those in Fig. 8.4, which shows typical records of maze learning.

Repetitive reinforcement eventually produces a maximum response beyond which we cannot show that learning has improved in terms of performance on any one trial. However, the longer we go on reinforcing a response beyond this maximum ('overtraining'), the more resistant it becomes to 'extinction', i.e. the animal will go on responding longer if reinforcement is removed.

Generalization and discrimination

If Pavlov conditioned a dog to salivate when a pure tone of, say, 1,000 cycles was sounded, it would also salivate but to a lesser extent when other tones were given. It *generalized* its responses to include stimuli similar to the conditioned one and the more similar they were the more the dog salivated. The opposite process to generalization is *discrimination*. Dogs naturally discriminate to some extent or they would salivate equally to all sounds, but their discrimination becomes refined after repeated trials when only one particular tone is followed by reward. We can accelerate discrimination if, as well as rewarding the right tone, we slightly punish the dog when it salivates to others. This 'conditioned discrimination' method has been of enormous value for measuring the sensory capacities of animals. After training to one particular stimulus—it may be a colour, brightness, shape, texture, sound, smell, weight, etc.—we then test to see how far the animal can discriminate this stimulus from others. We present it together with another stimulus of the same type and reward only responses to the former, perhaps giving slight punishment for incorrect responses. The two stimuli are made increasingly similar until there comes a point beyond which the animal can no longer learn to discriminate between them. This marks the limit of its sensory capacities as measured by its behaviour.

To give but three examples from many hundreds, this method was used

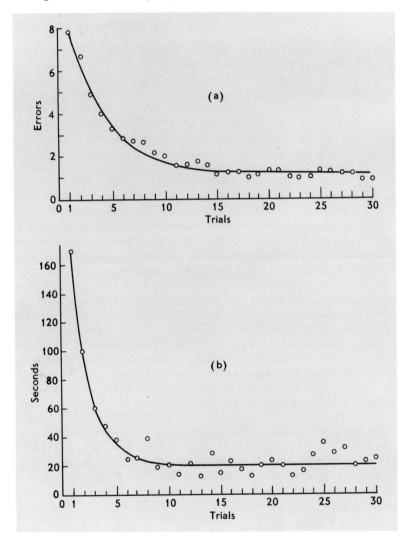

Fig. 8.4 The learning curves of rats learning a multiple T maze plotted (a) by errors, and (b) by time to reach the goal box. Each circle represents the mean value for 47 animals. (From Woodrow,[286] 1942, *Psychol. Bull.*, **39**, 1.)

by von Frisch[77] in his classical studies of the colour vision of bees; it was also used to examine the touch sensitivity of the octopus,[281] and the chemical senses of fish.[41]

Reinforcement

The concept of reinforcement has been a central one for the learning theories of Hull and Skinner and some objections to its universal application have already been discussed under 'contiguity'. Hull considered that reinforcement was essential for learning and that it consisted of a reduction in 'drive' or 'need'. A hungry animal reacts to a stimulus or performs an activity because in the past they have been associated with a reduction in its hunger. If this reward is removed, the learnt response gradually extinguishes.

This argument is sometimes used in reverse; the occurrence of learning in a particular situation is taken as evidence of drive reduction. Miller and Kessen[191] find that hungry rats learn to visit the alleyway of a maze where they can drink sweetened milk, in preference to a similar alleyway where milk is directly injected into their stomachs. However, they will learn to visit the latter if the alternative alleyways have no food. Miller and Kessen assume that hunger drive is reduced more by the act of drinking milk than by having it placed in the stomach, but that this also reduces drive. Similarly Myer and White[198] conclude that some rats have a spontaneous aggressive drive, because they will learn a simple maze with the opportunity to kill a mouse for a reward. We discussed other examples relating learning to aggressive motivation in Chapter 4, p. 81.

In its simplest form the drive-reduction hypothesis runs into trouble with 'avoidance conditioning'. Animals will learn to avoid punishment if it is signalled by some stimulus. A rat runs off an electrified grid when a light flashes once it has learnt that this signals that a shock will follow in 5 seconds. It will continue to respond in this way indefinitely and never gets another shock, hence the name 'avoidance conditioning'.

Here, on the face of it, is the persistence of a learnt response in the absence of any reinforcement, i.e. shock. Hullians can get round this difficulty by proposing that the initial shocks induced a state of 'anxiety' which continues to be aroused each time the rat sees the light. It is the reduction of this anxiety or 'secondary drive' which is rewarding. But if learning always entails drive reduction, we might repeat a question asked in Chapter 4, how many drives are there? Monkeys will learn to press a bar which moves aside a shutter for a few seconds and allows them to watch an electric train working! Do we ascribe this learning to a reduction in the 'curiosity drive'? As we shall see, the drive-reduction hypothesis has further trouble in accounting for learning during the exploration of a new environment (so-called 'latent learning') and for imprinting. There is a good discussion of this problem as a whole in Watson.[274]

Although we may reject the idea that a 'conventional' reward is necessary for all learning, we must remember that its effectiveness for simple associative learning is not in doubt. Loucks[162] stimulated a part of the motor cortex in the brain of dogs and thereby induced them to move a leg. He paired this stimulus with the sound of a buzzer, for more than 600 trials in some cases, without getting any trace of a leg movement to the sound given alone. Clearly, in this situation, mere association of the two events was inadequate to produce learning. Loucks then gave a small food reward following the sound and the induced leg movement, and within 6 trials the dog moved its leg spontaneously to the buzzer alone.

Olds[202] has found that electrical stimulation in some areas of the brain is 'rewarding' in itself. Rats learn to press a bar which delivers a brief electrical pulse to one of a number of sites in the hypothalamus, for example. Sometimes this self-stimulation is so intensely rewarding that a rat will press over 7,000 times per hour until it falls asleep exhausted! Presumably brain sites which produce this effect when stimulated are some part of the mechanisms whereby normal rewards are registered in the brain (see p. 223).

Many studies have been made to determine how the strength of motivation affects learning. Up to a certain level, learning for a food reward may be improved and accelerated by increasing hunger, but there usually comes a point at which very high motivation begins to interfere with learning. In anthropomorphic terms the animal is so desperate to get at the reward that he cannot 'give his mind' to the problem at hand.

Extinction

If we cease to reward a conditioned response it becomes reduced and eventually disappears. It is much easier to extinguish classical CRs than instrumental CRs although we do not know why this should be so. After a rat has learnt to press a bar for food, the proportion of rewards to presses can be reduced to as low as 1 in 100 in some cases, and the rat goes on pressing. If rewards are stopped altogether it is a very long time before the response finally extinguishes.

Pavlov realized that an extinguished CR did not just disappear and leave the animal as it was before conditioning started. In the first place, if we simply leave the animal alone for a few hours and then give it the CS again, the CR returns, i.e. it shows ***spontaneous recovery***. This recovery is not back to the original level and the response extinguishes more rapidly, but this process of a pause followed by spontaneous recovery can be repeated several times.

A second way of reviving an extinguished response is to give a novel stimulus along with the CS. A dog which has had its conditioned salivary response to a bell extinguished, salivates again if a light flashes as the bell is sounded. Similar results have been obtained with rats in Skinner boxes. Pavlov called this process 'disinhibition' because he regarded extinction as another new learning process which inhibited the original CR. Neutral stimuli presented with the CS early in the original acquisition of the CR often 'inhibit' it temporarily and reduce its strength. Similarly, perhaps the neutral stimulus disinhibits an extinguished CR by inhibiting the new learning that takes place during extinction.

Having discussed some of the more important properties of associative learning, we can now complete our examination of Thorpe's learning classification. The remaining categories are certainly more complex and less clearly defined than the earlier ones.

Latent learning

Thorpe defines latent learning as '. . . the association of indifferent stimuli or situations without patent reward'. By this he implies that the actual process may involve associative learning, but of course the key distinction is the lack of any obvious reward or drive reduction and the fact that what is learnt may not be obvious at the time—it remains 'latent' or hidden. Latent learning in its natural setting will often result from an animal exploring its new surroundings. Suppose we put a maze up against a rat's home cage so that he can wander into it at will. The rat is not hungry or thirsty, nor does he find any food or water in the maze, but he explores through it, sniffing into corners, running down blind alleys, retracing his steps and so on. Does he learn anything about the maze as he explores? According to strict Hullian theory he cannot, because he has received no reinforcement for doing so. Tolman's theory, on the other hand, lays less emphasis on reinforcement and predicts that learning would occur. The test comes when we train our rat to run through the maze when hungry to get a food reward at the other end. Does he perform better than a rat put hungry into the maze for the first time? If so then his previous learning was 'latent', in that it was not revealed until we gave him a chance to put it to good use in finding food. Munn[197] (Chapter 9) provides an excellent discussion of the dispute between the Hullians and the Tolmanians which continued for some years in the experimental psychology journals.

To a biologist the whole controversy seems quite unreal, because it is so obvious that animals do learn about their environment when they explore. A detailed knowledge of the geography of their home area will often mean the difference between life and death to a small mammal or bird as a predator swoops down. Barnett[13,14] discusses the nature of exploratory behaviour and concludes that it can be distinguished from the appetitive behaviour of a specific motivational state. There are times when the animal deliberately seeks out new stimuli and 'explores' in the true sense. Informa-tion it gains about its environment in this way can later be employed when it is searching for food or a mate, for example. Some workers have proposed that there is an 'exploratory drive' (which might be reduced during learn-ing) but there is no good evidence that exploratory behaviour is organized in the same way as hunger and thirst. For example, Halliday[94A] found that a rat's tendency to explore a new situation is not reduced if it has just been exploring something else—if anything, its exploration increases.

We know that some birds and mammals 'explore' in the sense outlined above. In the other groups all we know for certain is that animals learn, often in fantastic detail, the geography of their home area. Many insects make special 'orientation flights' during which they make a 'fix' of the home area's position relative to the sun and landmarks near by. If a honey-bee colony is shut up in the hive and moved to a new site, a large proportion of the workers make orientation flights when they first leave the hive in its new position. They hover outside the entrance hole and then circle, gradually increasing their distance from it before flying off. During this orientation flight, lasting only 1 or 2 minutes, they learn the new position in sufficient detail to be able to return from long foraging flights. Many of the Hymenoptera have this amazing orientation ability because they build nests which require repeated attention and the insect needs to return many times to precisely the same spot. It might be possible to fit learning of this type into a drive-reduction framework, but there seems little profit in the exercise. This ability has been built into the insect's nervous system by natural selection and we shall need to study it as a problem in ontogeny and evolution as well as learning. Thorpe gives numerous examples of similar orientating ability in animals as diverse as limpets, fish and newts.

Insight learning

Insight appears to us as the highest form of learning. Everyone can recall occasions when the solution to a problem has 'come in a flash', perhaps as the climax to several minutes of concentrated thinking. It is obviously difficult to demonstrate conclusively that there are similar pro-cesses going on in animals. Most workers have used the term 'insight' when, for instance, they observe animals solving problems very rapidly, too rapidly for normal trial and error. At least, too rapidly for the animal to carry out actual trials, but there is the possibility that it is 'thinking' about them and trying them out in its brain. This would imply that the animal can form ideas and 'reason', and studies on animal reasoning seem doubt-fully distinct from those on insight.

Maier (see Maier and Schneirla[173]) defines reasoning as '. . . the ability to combine spontaneously two or more separate or isolated experiences to form a new experience, which is effective for obtaining a desired end'.

Maier and others have tested for the existence of reasoning in the rat using a number of methods, mostly involving the animal making detours or taking short cuts through mazes. Figures 8.5 and 8.6 with their captions provide two typical examples, and there are two general features of such experiments. Firstly, they all include a period of exploration prior to testing which will mean that final performance depends to a great extent on latent learning. Secondly, we judge performance largely by speed of solution: if the animal makes mistakes and has to explore further we assume it is using a simpler, trial and error form of learning.

Fig. 8.5 One type of apparatus used by Maier to test 'reasoning' in the rat. The pathways are 8 feet long and the small tables vary in size, shape, and character. S_1, S_2 and S_3 are wooden screens placed on the tables to obstruct vision from one to the other. After exploring the three tables and runways, the rat is fed, let us say, on table A. It is then, let us also assume, placed on table C. After reaching the joint origin of the three paths, the animal now has a choice between A and B. If it chooses A, it is credited with a correct response. Exploration precedes each test and the rat is started from different tables from test to test. In a group of such tests, a score of 50% would occur by chance; some rats score much better than this. (From Maier,[172] 1932, *J. Comp. Neurol.*, **56**, 179.)

It is perhaps unwise to read too much into ordinary detour experiments because some insects show extraordinary ability to handle this type of problem. Thus Hebb and Williams[101] suggest that a good measure of 'intelligence' is obtained by using a series of movable barriers on an open space with a fixed starting-point and goal. Once it has become thoroughly accustomed to the space and goal the barriers are moved between each trial to present the animal with a series of simple detour problems (see Fig. 8.7). Scores can be given for directness of runs between start and goal box. Modifications of this technique have been extensively used by experimental psychologists. Although of course it has not been tried, there seems little doubt on the preliminary evidence given by Thorpe,[253] that the wasp *Ammophila* would score as highly as a dog on this test! In the natural situation, *Ammophila*'s starting-point is the place where it has stung a caterpillar, the goal is its nest to which it drags the prey. It has never made

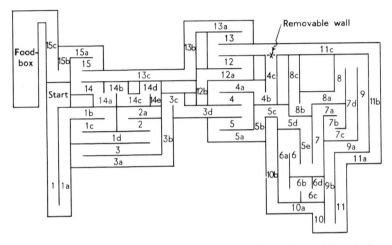

Fig. 8.6 A complex maze used by Shepard to test 'reasoning'. After rats have learned the maze, the section indicated by X is removed, thereby causing a previous blind alley to become a short cut. Having discovered the change whilst running along 11c, and exploring from there a little into 4c, some rats entered 4 (and thence 4a, 4b, 4c) instead of 5 on the next trial. (From Maier and Schneirla,[173] 1935, *Principles of Animal Psychology.* © 1935. By permission of McGraw-Hill, New York and Maidenhead.)

the journey on land before, although it has made orientation flights around the nest. If whilst it is moving, barriers are placed in its path, it usually changes direction *instantly*, so as to carry itself smoothly around the barrier with the least effort. *Ammophila*'s remarkable ability to detour is probably based on orientation with respect to the sun, and one would hesitate to call it 'insight'. Even so it must make us reconsider our judgment of how 'insightful' are the detour abilities of higher animals.

The classic example of insight in animals came initially from the work of Köhler[140] on chimpanzees. Presented with a bunch of bananas too high to reach, they would pile up boxes to make a stand for themselves or fit two sticks together in order to pull down the bananas. Often they arrived at this solution quite suddenly, although they benefited by previous experience of playing with boxes and sticks (latent learning) and showed considerable trial and error when actually building a stable pile of boxes.

Köhler's chimpanzees were using knowledge obtained in one context (something of the properties of sticks and boxes) and applying it in another. There is no question but that apes and other primates can show true reasoning on occasions. Many dog owners will cite examples of their pets doing likewise; this is possible, but we must take care to exclude other explanations. Thorpe[255] discusses some other types of behaviour in birds and

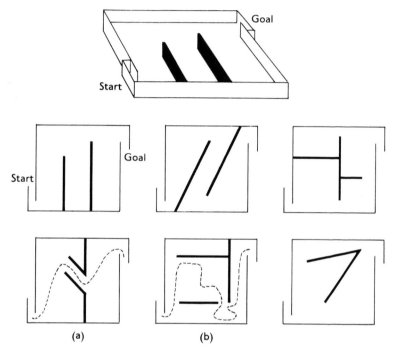

Fig. 8.7 One type of situation suggested for measuring animal 'intelligence'. A simple enclosed arena, illustrated at the top, has moveable barriers. The diagrams below show a series of different detour problems with which the animal can be presented in turn. The route shown on (a) would rate a high mark, that on (b) a poor one. The marks from a number of problems are averaged to give an 'intelligence score'. (From Hebb and Williams,[101] 1946, *J. gen. Psychol.*, **34**, 59.)

mammals which are fair evidence of reasoning, and the reader is referred to Chapters 6 and 7 of his book.

Imprinting

Although some argue otherwise (notably Hess[105]), probably the majority of workers in the field would now agree that it is incorrect to consider imprinting as a separate category of learning. Imprinting it is often said, is characterized by its context and not its content. However because it often concerns learning in young animals at a highly sensitive period in their development, when they are receiving their very first stimulation from the outside world, imprinting does comprise a very different body of observations from that provided by more conventional types of learning. It has also had a very different history from that of learning generally. The first to make systematic observations was Spalding[249] who found that young chicks, 2 or 3 days from hatching, would follow any moving object. It was an ethologist, Lorenz,[155] who introduced the topic to most behaviour workers. Lorenz described imprinting as a unique learning process whereby the young of precocial birds (i.e. those that can walk at hatching and do not stay in the nest) form an attachment with a 'mother figure'. Normally, of course, this is their real mother, but they will become attached to almost any moving object and subsequently treat it as a mother. Lorenz considered that imprinting, unlike other forms of learning, was irreversible, was restricted to a brief 'sensitive period' just after hatching, and that a young bird's choice of a mother also affected its choice of a sexual partner when it became mature.

Imprinting has aroused great interest both among psychologists and ethologists. Most of the work has been, as originally, with birds and we can concentrate our attention on them, but the importance of imprinting-like processes in insects and mammals is recognized. Sluckin[245] and Bateson[21] provide excellent surveys of the whole subject.

Although imprinting seems superficially so different from other forms of learning, it can be investigated in much the same way; i.e. a period of training followed by a test of retention and discrimination. For example, in one of his tests Guiton[90] used a circular runway around which an object could be moved suspended off the ground. Chicks were put into the runway singly when 1 to 2 days old and exposed for 30 minutes to one of two moving objects, a red cardboard box or a green box, which differed in shape and size. The chicks usually followed their box for a good proportion of the time. Three days later, they were put back in the runway at a point equidistant between both boxes, which were then set in motion. Guiton recorded how much time was spent following each box or standing close to it, and found that after a single exposure the chicks discriminated almost completely, and ran at once to the model with which they had previously been trained. In this, as in a number of other studies of a similar type, the fact that the chicks follow the model is used as a convenient measure for the strength of the response. But following is not an essential part of imprinting —the chicks revealed their imprinting by their choice of the model to which they had been previously exposed.

We can organize our brief discussion of imprinting by considering four questions:

1. What are the stimuli which direct the young birds' approach?
2. Is there a sensitive period and, if so, what ends it?
3. What are the results in adult life of imprinting during infancy?
4. What is the relationship of imprinting to other forms of learning?

1. WHAT ARE THE STIMULI WHICH DIRECT APPROACH? Lorenz's early observations were most dramatic because he got broods of greylag goslings to imprint on himself. There seems to be no limit to the range of visual stimuli for imprinting. Birds have been imprinted upon large canvas 'hides' inside which a man can move, down through cardboard cubes and toy balloons to matchboxes. Colour and shape seem to be equally immaterial. Nor is movement necessary; a stationary object will attract young birds provided it contrasts with its background, so will flashing lights. Bateson and Reese [21A, 21B] have shown that within a few minutes day-old chicks and ducklings will learn to stand on a pedal in order to switch on a flashing light, which they then approach. The rapidity with which they acquire this response suggests that the light has reinforcing properties even before the young birds are imprinted to it. They actively seek such stimulation and such attraction forms the basis for the imprinting attachment.

Auditory stimuli are also attractive to young birds and will often enhance imprinting to objects close to their source. Sound is almost essential to induce following in mallard ducklings, for example. In some cases the response to sound goes beyond this and Klopfer[138] has demonstrated that young wood-ducks show a form of auditory imprinting, moving towards a sound source. They will respond to a wide range of sounds on the first exposure but subsequently discriminate against unfamiliar sounds. Wood-ducks nest in holes in trees, and normally the young hear their mother calling from the water outside the nest hole before they have ever seen her.

2. IS THERE A SENSITIVE PERIOD AND, IF SO, WHAT ENDS IT? By using the term 'sensitive period' we imply that only within this period will exposure to an object result in the bird becoming attached ot it; 'critical period' is often used synonymously. The sharpness of the sensitive period depends to some extent on how exacting are the criteria adopted. If the criterion 'follows a moving object at first exposure' is used, then many precocial birds will satisfy this for 10 days or more after hatching. However, if one takes as a criterion 'the formation of a lasting attachment following a single exposure to an object', then the sensitive period may appear much more limited. Figure 8.8 illustrates some results using both types of criteria. Those from Ramsay and Hess[210] measured how well mallard ducklings discriminate and stay close to a model, 5 to 70 hours after a single 30-minute exposure to it. Boyd and Fabricius[34] simply scored birds which followed a model on their initial exposure.

Guiton[90] found that chicks kept in groups ceased to follow moving objects 3 days after hatching, but chicks reared in isolation remained responsive much longer. He could show that the socially-reared chicks became imprinted upon one another. This may be important in the natural situation. Boyd and Fabricius[34] point out that mallard ducklings do not

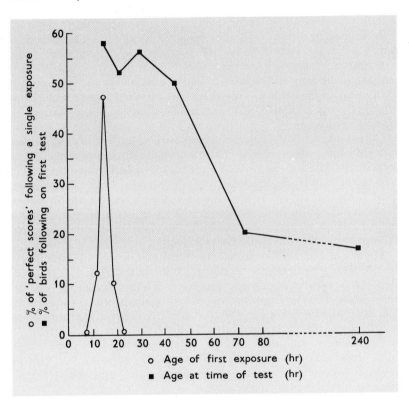

Fig. 8.8 Two ways of expressing the sensitive period for imprinting. The open circles show the very sharp peak obtained when mallard ducklings were tested for discrimination and following some time after a single training exposure at the ages shown. The black squares show the much broader peak obtained by scoring simply the percentage of birds which follow a moving object on the first exposure. (Data from Ramsay and Hess,[210] and Boyd and Fabricius.[34])

normally leave the nest until the second day after hatching, well past the peak of the sensitive period. By this time some of them may be, at least partially, imprinted upon each other. Even if only a few of the brood actively approach and follow the mother bird as she leads them to water, the brood will act as a group and stay together because the rest will follow the maternally imprinted ducklings.

Since social rearing restricts the sensitive period it may be that the imprinting process itself plays a part in ending this period. Bateson[20] has shown that 3-day-old isolated chicks prefer to follow a model whose pattern resembles that on the walls of the pen in which they have been reared. Similarly Taylor *et al.*[251D] found that exposure of young chicks to coloured walls in their pen shifts their subsequent choice of colours towards that which is familiar, even if the initial exposure was only 15 minutes. Such experiments strongly suggest that imprinting involves the young animal learning to discriminate the familiar from the unfamiliar and approaching the former. In the absence of any conspicuous moving object the chick, in effect, becomes imprinted on its surroundings and its responsiveness to new objects declines.

Other workers have laid more emphasis on the growth of fear itself as the cause of waning responsiveness. There is no doubt that escape responses do increase from the first day; newly-hatched chicks or ducklings will approach new objects, but by the third day they will flee and crouch. The point at issue is whether the escape tendency only grows when the mother figure becomes sufficiently familiar through imprinting to make other environmental stimuli recognizably unfamiliar, or whether it grows anyway, independently of such experience, as a result of maturation.

Moltz and Stettner[192] reduced the visual experience of ducklings by rearing them with translucent hoods which meant that their eyes received light but no pattern. Such birds when tested at 2 or 3 days of age without their hoods, followed a moving box more than control birds, and showed less fear. This certainly suggests that previous visual experience of the surroundings does affect the growth of fear.

Loud noises or mild electric shocks, which would be frightening to older birds, make chicks follow their mother figures all the harder. This is probably because heightened arousal leads to a more intense response to both familiar and unfamiliar objects.

3. WHAT ARE THE SUBSEQUENT RESULTS OF IMPRINTING?   Lorenz and others observed that young birds reared by a foster mother of another species ceased to follow her as they became independent, but later courted and attempted to mate with birds of the foster mother's species when they became sexually mature.

These observations have been amply confirmed by recent experiments; both male and female turkeys have been sexually imprinted on humans[230] and cockerels imprinted onto cardboard boxes which they courted and attempted to mount.[91] Systematic cross-fostering experiments designed to study sexual imprinting have been carried out both with various species of ducks and of Estrildine finches.

With the finches, Immelman[129] placed a single egg of one species e.g. zebra finch, in a clutch of a second, e.g. Bengalese finch, and allowed the Bengalese parents to rear the whole brood. Subsequently, cross-fostered zebra finch males were isolated until they were sexually mature. Immelman then gave them a choice between a zebra female and a Bengalese. The results were quite unequivocal; the zebra male directed all his courtship towards the Bengalese female. This preference was all the more striking because when a zebra male was put in with the two females the zebra female usually responded at once with all the usual con-specific greeting calls. The Bengalese female was, at best, neutral and usually showed avoidance as he approached her.

Immelman found that if cross-fostered males were obligatorily paired with females of their own species most would eventually mate with them and raise broods of young. Astonishingly, such experience did *nothing* to alter the preference described above. When once again given a choice, the males totally-ignored con-specific females and courted the foster species.

Immelman has not yet reported on the behaviour of cross-fostered female finches; it is not so easy to obtain an objective measure of their choice of a mate because the female's rôle in courtship is more passive. In his duck experiments Schutz[234] found that only males imprinted on the foster-species and he suggested that this is because whilst each duck species has a distinctive male plumage pattern, the females are all much alike and cryptically coloured. It may be 'easier' to evolve an inherited discrimination for a particular male pattern than for a female (see p. 45 for an analogous example from fish). Supporting this conclusion, Schutz found that in the Chilean teal where both sexes are alike and have cryptic plumage, the female also imprints. However, this does not explain why geese, which resemble the Chilean teal in this respect of plumage, do not show sexual imprinting in either sex. Nor does it explain why male shelduck, whose females are brightly coloured like themselves, do imprint upon females of other species.

In general the effects of cross-fostering were not so strong in the ducks as in the finches, at least in the sense that not all cross-fostered males were imprinted. Thus of 34 mallard males reared with other species (either foster siblings or a foster parent) 22 subsequently tried to pair with the foster species. They paired with females if they had been reared with a foster mother, but formed homosexual pairs if the foster parent had been a male. In nature, imprinting upon the mother as a figure to be followed and sexual imprinting normally go together, but we do not yet know whether they are identical processes. The sensitive period for sexual imprinting extends far beyond that for the following response. Schutz reared some ducklings with their own species for up to 3 weeks and then cross-fostered them. One-third of such birds still became sexually imprinted on their foster species, although this is to be compared with the two-thirds which responded when cross-fostered from hatching. Immelman's evidence suggests that sexual imprinting is complete in the finches by 33 days of age, i.e. when fledglings are just able to look after themselves. Keeping males with the foster species for 94 days, i.e. until just before sexual maturity, had no extra effect.

The strength of imprinting in the finches was not affected by their experience after leaving the nest—as we have seen no amount of subsequent contact with their own species modified their preference. It is unlikely that sexual imprinting is so strong in all birds, but we lack much comparative evidence. Imprinting of this type certainly occurs in pigeons but note that it *cannot* occur in the European cuckoo and other parasitic birds whose young are always reared by foster species.

The experiments just described reveal the critical nature of early experience in determining the young animals choice of a sexual partner. Both Immelman and Schutz point out that, dramatic though the results of cross-fostering may be, it is still easier to imprint a male on his own species than a foster species. Probably there are certain inherited tendencies which normally strengthen attachment to the con-specific female in the natural situation, but as yet we know very little about the interaction between inherited and learnt responsiveness towards sexual partners.

4. WHAT IS THE RELATIONSHIP OF IMPRINTING TO OTHER FORMS OF LEARNING?   As we mentioned at the beginning of our discussion of imprinting, the conditions under which it occurs are very different from those of other learning and this has led to claims that it is unique. As knowledge accumulates this conclusion seems to be harder to maintain. Imprinting, like associative learning, improves with repeated trials and imprinted animals show the normal type of discrimination and generalization about their imprinting object. Amongst its supposedly unique features, its speed, irreversibility, sensitive period and lack of reward are most quoted. Other types of learning, such as orientation flights of bees, are equally rapid. Imprinting is not necessarily irreversible although, as we have seen, the fact of being imprinted certainly does interfere with subsequent learning—perhaps because of the escape tendency evoked by unfamiliar objects—and may thereby terminate the sensitive period. On p. 34 we considered song learning in the white-crowned sparrow, and here too the sensitive period ends in a male's first spring after he has copied the songs of nearby adults. There is certainly no obvious reinforcement for imprinting. It starts with an inherited tendency to approach conspicuous features in the environment and, although a natural mother will provide

plenty of reinforcement by her warmth and shelter, these are not necessary for imprinting to occur. However, now that the extent of latent learning is recognized, the absence of reinforcement cannot be considered unique.

There is no point in trying to fit imprinting into the mould of associative learning; it clearly is rather different, but it will be just as profitable to look for similarities as to emphasize differences.

Imprinting in mammals

Imprinting-like phenomena are clearly involved in the social development of mammals. It is common knowledge that the younger we take and rear a litter of wild mammals the easier it is to tame them. Orphan lambs reared by humans follow them about and often show little attraction towards other sheep. This is not just a filial attachment, a form of sexual imprinting must also occur and zoo authorities know to their cost that hand reared animals are often useless for breeding when they are mature.

The behaviour of mammals is very much dominated by their sense of smell, and it is perhaps not surprising to find that their early olfactory experience often affects their choice of a mate. In mice, rats and guinea-pigs, mature animals are more attracted by others whose scent matches that which was present in the nest during the time they were being reared (see Carter and Marr[44B]).

The development of social responses is also highly dependent on early experience. Scott and Fuller[236] summarize the extensive work of their group on dogs (see also Scott[235]). They have found that there is a sensitive period from about 3 to 10 weeks of age during which a puppy is forming normal social contacts. If isolated beyond the age of 14 weeks they no longer respond, and their behaviour is very abnormal. A very short exposure at the height of the sensitive period is sufficient for them to form a normal relation with human beings. Dogs, like some sexually imprinted birds, seem perfectly capable of accepting both man and their own species as social partners.

There is some dispute over the recognition of sensitive periods in the social development of mammals. (Papers relating to this dispute are introduced and collected together in McGill,[168A] see also Denenberg.[61B]) As we have seen when discussing sensitive periods for imprinting in young birds, it is often very difficult to distinguish between a change in responsiveness that occurs as a result of previous experience and a maturational change that occurs irrespective of what has gone before.

The study of sensitive periods nearly always involves either isolating the young animal from supposed influences, or alternatively attempting to add extra experience at different periods, (as in the 'early handling' experiments discussed in Chapter 2, p. 24). In either case the young mammal is developing in a changed environment and this often means there are loopholes in any argument about sensitive periods.

To some extent this argument is merely academic, but it relates to one's view of behavioural development. Is it a totally continuous process with interactions between the growing animal and its environment possible at every stage? Or is it more like embryonic development where we know that certain events must take place within a critical period if they are to take place at all? Once the critical period is past the embryonic cells are no longer competent to respond.

The development of behaviour probably exhibits both sets of characteristics and much will depend on a particular animal's life history. With an infancy period as brief as that of a zebra finch, which is sexually mature at 3 months, behavioural development is highly compressed and imprinting takes on an all-or-nothing, irreversible appearance. The infancy of some mammals is very prolonged and development appears much more continuous. The work of Harlow's[97] and Hinde's groups with rhesus monkeys has vividly shown how the acquisition of normal sexual and maternal behaviour depends on social contact with their mother and siblings. Maternal deprivation, even for a few days in the early stages has long-lasting effects.[112D]

Imprinting has always attracted the attention of psychiatrists because there is no doubt that the human infant is also extremely sensitive to maternal deprivation and perhaps particularly so between the ages of 18 months and 3 years. Bowlby[33] has argued the case for a critical period at this age when the infant forms a bond with its mother which later serves as a basis for all other social attachments. However the long childhood years allow far more chance of 'repair' than in short-lived animals.

Finally, we can mention one really clear example of a sensitive period, but this time in an adult mammal. Klopfer *et al.*[139] have shown that shortly after she gives birth, a mother goat has a brief period when she is sensitive to the smell of her kid. Unless she has access to the kid within an hour of birth it is subsequently rejected, but 5 minutes' contact during the sensitive period will prevent rejection, even after a separation of 3 hours.

THE COMPARATIVE STUDY OF LEARNING

Comparative psychology has a long history, although it has often been criticized with some justification—notably by Beach[23A] in a famous paper entitled 'The Snark was a Boojum'—for abandoning its comparative aims and concentrating almost entirely on two animals, the white rat and the pigeon. A far wider range of animals is now under scrutiny and we might hope that a comparative survey of learning capacities would yield information both on the evolution of learning and on the generality of any 'laws of learning' which have been put forward.

One approach has been to examine the correlation between brain development and learning ability. Dethier and Stellar[64] include an excellent introductory survey of nervous system structure through the various animal groups, and they discuss how far behavioural complexity can be linked with brain development. The prevalence of learning increases through the vertebrate series and we can roughly equate this increase with the evolution of a large brain. It is not, of course, just brain size that is important; whales and elephants have larger brains than men but smaller, though considerable, learning powers. It is less easy to construct a series among the invertebrates, which are much more heterogeneous, but higher insects and cephalopods have the largest brains of their respective phyla, Arthropoda and Mollusca, and also the greatest capacity for learning.

Again, within the vertebrates the most dramatic aspect of brain evolution is the growth of the cerebral hemispheres, especially their cortex, which reaches a climax in the primates. However, we must avoid any simple equation of cerebral cortex with learning ability. The birds have in the past often been underestimated because, although their brains are relatively large, those parts homologous to the mammalian cerebral cortex are small. But birds have evolved along a line separate from the mammals for over 200 million years. They have evolved another type of brain structure, and their learning ability is in some respects second only to the primates. A converse example has already been mentioned on p. 18 for amongst the insects the Diptera and Hymenoptera have similar brain development but very different degrees of learning ability.

It is clear that brain structure alone is inadequate as a guide to learning abilities, and to study the evolution of learning we need to compare how different animals perform on particular behavioural tests. We are immediately faced with a number of problems. Some concern the selection of representative animal types for our evolutionary analysis. It is all too easy, to refer to 'higher' and 'lower' animals, and within the vertebrates, for example, we often find the sequence fish, amphibian, reptile, bird, mammal quoted as an evolutionary scale of increasing complexity of behaviour and increasing learning capacity.

The construction of such a scale from living representatives of each class, ignores the actual course of evolution. We have just mentioned the completely separate histories of the birds and the mammals. All the living vertebrates are equally distant in time from their common ancestors and all are specialized for their particular mode of life. We should be naïve to expect the learning capacities of a modern teleost fish (e.g. a goldfish, commonly used in learning studies) to reflect accurately those of the ancestral fish from which the teleosts and other vertebrates diverged some 400

million years ago. Hence the construction of a valid phylogeny of behaviour is fraught with difficulties because behaviour does not fossilize.
Hodos and Campbell[117A] provide a vigorous critique of the phylogenies
that have sometimes been constructed by comparative psychologists.

A further problem for comparative learning studies is that of devising
truly comparable situations for testing different animals which vary so
widely in their sensory capacities and manipulative ability. The procedures
needed to measure discriminative conditioning in an octopus, a honey-bee
and a rat have to be very different and we can no longer be sure that
problems are of equal difficulty or that the animals 'see' them in the same
way. Motivation and reinforcement present further problems; we have
seen (p. 189) that the level of motivation affects rate of learning and may,
indeed, determine whether the animal learns at all. How can we equate level
of hunger motivation in, say, a rat and a fish? The latter may live for weeks
without food, the former only days. It is just as difficult to equate reinforcement between different animals. A small piece of food may be an excellent
reinforcement for a hungry mammal, but mean much less to a fish and less
still to an annelid. It is perhaps easier to equate punishment, since all
animals 'dislike' electric shocks. Even here there are difficulties, because
shock or fear affects the behaviour of animals in such diverse ways. As we
discussed earlier (see p. 185) animals come to learning situations with a
good deal of built in bias. To equate the effects of punishment we require
some knowledge of the animal's natural responses in fearful situations. For
example, avoidance conditioning apparatus (see p. 188) usually requires an
animal to run off an electrified grid when shocked, and subsequently to
learn that a buzzer or a light flash signals that the shock is coming. However, some rodents do not readily run when alarmed, they freeze and may
stay motionless for long periods even when receiving electric shocks. Such
an animal will appear very slow to learn when compared to one that runs
easily. From the experimenter's point of view active animals are usually
much more amenable in almost any type of learning situation (see Denny
and Ratner[61D] for a good review of these problems).

Nevertheless with some ingenuity and much caution, it is possible to
devise test situations which probably give diverse types of animals a fair
chance to demonstrate their abilities. For example, Bitterman[27] describes
'Skinner boxes' for fish and turtles which provide food rewards; Schneirla
(see Maier and Schneirla[173]) and Vowles[270] both found that the best
reinforcement for maze learning in ants was to get back to their nest. For
newts[268] and some annelids,[52] the best reward was a brief period in which
the unwilling subjects were simply left in peace!

Despite the problems involved in its study, the evolution of learning
remains a fascinating topic. It is reasonable to examine our classification
of learning to see whether the different categories have any phylogenetic

validity, i.e. can they be ranked in order of appearance on the evolutionary scale, at least as far as this scale can be represented by living examples?

Undoubtedly the oldest type of learning is habituation which, as we discussed earlier, is found in all animals. It is easy to understand the selective advantage it confers because habituation saves animals wasting time on stimuli which are unimportant, and this will be advantageous no matter how limited is their repertoire of responses.

Some form of associative learning must have been the next step in evolution, but it appears to be a large one. Associative learning does share certain features with habituation—for any type of learning to occur the animal must be able to distinguish familiar from unfamiliar situations—but apart from this it is hard to construct evolutionary intermediates. However the phenomenon usually called 'sensitization' may represent a stage in the evolution of associative learning and is, in some ways, the opposite of habituation. We speak of sensitization when an animal's responsiveness to a variety of stimuli is increased following a reward or punishment. For example. Evans[70B] found that the worm *Nereis* (see p. 177) will occasionally emerge from its tube when a light is flashed, and in one experiment 21% of worms did so. However following a single food reinforcement—given in the absence of light—more than 60% of the worms emerged to the next flash given 30 minutes later. Similarly an octopus which has just been fed is far more likely to attack cardboard shapes suspended in its tank than is an unfed animal. Sensitization works even more effectively with punishment. Following electric shock to their feet, both cockroaches and rats will flee from a variety of novel stimuli which have never been associated with shock and which do not elicit escape from unsensitized animals.

To evolve true associative learning the animal must begin to discriminate between stimuli that occur close to the reinforcement and those that do not. Clearly a heightened responsiveness at the time of reinforcement will help it to do this. Unequivocal associative learning has been shown for the arthropods and the cephalopod molluscs (the octopus and its relatives) among the invertebrates. Results that seem to indicate such learning in annelids and flatworms are—so far—almost all susceptible to other explanations and often sensitization is one such (see Evans,[70B] and Thorpe and Davenport[256] for reviews of the problems of interpretation here).

Amongst the vertebrates, we might hope by a good series of learning tests to be able to record stages in the evolution of 'intelligence', and speed of learning—a quality much admired in human school children—might seem to be one useful measure. However a brief survey of the literature shows that speed alone does not tell us much. Ants and rats, for example, show very comparable speed when first learning a fairly complex maze. Gellerman[82] describes in detail experiments in which two chimpanzees

and two 2-year-old children were learning that a food reward was associated with a white triangle on a black square and not with a plain black
square. One child learnt in a single trial, but the other took 200 trials and
both the chimpanzees took over 800 trials to reach the criterion of 19 correct trials out of 20. On a comparable test most rats would learn in 20 to 60
trials, though admittedly they would usually be mildly punished for wrong
responses as well as rewarded for correct ones. Discrepancies of this kind
abound both within and between species and we have no reliable evidence
that speed of initial learning for simple associative problems varies within
the vertebrates, or even between them and the higher invertebrates.

However, speed is clearly only one aspect of learning; we might also ask
what is learnt. For instance, Fig. 8.9 shows that, although the chimpanzee
may take longer to learn the triangle discrimination outlined above, he
learns more about 'triangularity' than does the rat. One aspect of 'intelligence' is the ability to strike a reasonable balance between generalization

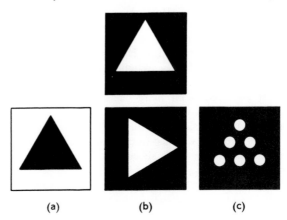

(a)   (b)   (c)

Fig. 8.9 The concept of 'triangularity'. Trained to respond to the top figure, a
rat makes random responses to any of the lower figures. A chimpanzee responds
to (a) and (b), but makes random responses to (c). A 2-year-old human child
recognizes 'a triangle' in (a), (b) and (c). (From Hebb,[100] 1958, in *Biological and
Biochemical Bases of Behaviour*, H. F. Harlow and C. N. Woolsey, eds. University
of Wisconsin Press, Madison. By permission of the Regents of the University of
Wisconsin.)

and discrimination in tests of this type. Similarly, if we consider more
complex forms of learning then we can at least note some gradation of
ability within the vertebrates.

Testing for **learning sets** has been useful in this respect. If an animal
can form a learning set it means that it can learn not just a problem, but the
principle behind it and can steadily increase its learning speed when given

a series of similar problems. Harlow[95] has described the basic technique with primates.

A monkey is presented with a pair of dissimilar objects—a matchbox and an egg-cup, for example. The matchbox, no matter where it is placed, always covers a small food reward; the egg-cup never has a reward. After a number of trials the monkey picks up the matchbox straight away. Now the objects are changed, a child's building block is rewarded, a half-tennis-ball unrewarded. The monkey takes about the same time to learn this; again the objects are changed, and so on. After some dozens of such discrimination tests the monkey learns each discrimination much more rapidly, though viewed as an individual problem it is just as difficult as the first one. Eventually after 100 or so tests the monkey, presented with a pair of objects lifts one, if it yields a reward he chooses it for all subsequent trials, if it is unrewarded he chooses the other (rewarded) object on the next and all succeeding trials. He has learnt the principle of the problem or, in Harlow's terminology, he has formed a 'learning set'.

This is one type of learning set based on successive trials of discrimination. Perhaps a simpler version of the same type is the 'repeated reversal' problem. Here we train the animal to select object A in preference to object B. Once learnt, object B is now rewarded and A unrewarded; when this first reversal is learnt, the reward is again given with A, and so on. If the animal gets progressively quicker at learning each reversal this again implies it has learnt a principle.

The ability to form learning sets was once regarded as a property of the higher mammals only, but we now know this is not the case. Warren[273] summarizes comparative data from fish, turtles, birds, rats, cats and primates. Although further research may prove otherwise, it looks as if this criterion does separate the fish, which cannot form learning sets, from all the other vertebrates. Recently Mackintosh[169] has provided good evidence that the octopus can form repeated reversal sets, so in this respect the ability of a higher invertebrate exceeds that of a lower vertebrate. No insect has yet been tested in this way.

Although speed of learning a simple discrimination does not vary much between groups, the speed with which learning sets are acquired does change dramatically. Figure 8.10 shows how much faster a monkey acquires a set than a rat, which shows no improvement at all for some 800 successive problems but thereafter slowly improves.

With learning sets, and with more complex problems, the superiority of primates is undoubted, but it seems to be a quantitative superiority. We have no evidence that they possess any abilities which are not foreshadowed earlier in the mammalian series.

One cannot help feeling vaguely dissatisfied by this conclusion. The rather circumscribed artificiality of many learning experiments does not seem to do justice to the incredible flexibility and ingenuity of the higher primates, such as chimpanzees. Here we are undoubtedly influenced by their similarity to ourselves and particularly because they can manipulate objects as we do. We may underestimate the intelligence of other animals because they lack good hands and good eyesight. Only recently have we become aware of the remarkable intelligence of dolphins because their structure and environment is so different from our own.

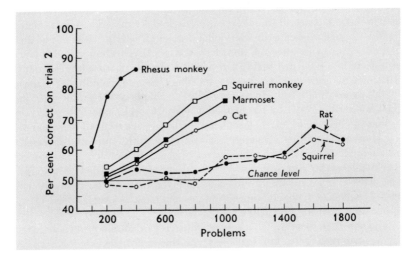

Fig. 8.10 The rate at which various mammals can form discrimination learning sets. With each new problem the animal's choice on the first trial has to be random, but if it has learnt the principle behind the problems, trial 2 should be correct. Note how long it is before the scores of rats or squirrels on trial 2 become better than chance or 50%. Many monkeys reach almost 100% within 400 problems. (From Warren,[273] 1965, *The Behavior of Non-human Primates*, vol. 1. Academic Press, New York and London.)

Studies on the behaviour of chimpanzees in the field have revealed that they can select and fashion simple tools. Goodall[84] has described how they will chew up leaves to use as a sponge for getting water out of crevices. They also select twigs, and peel off their side branches until they make a probe of a size suitable for extracting termites from their nests. Such observations should not, perhaps, surprise us because it now seems certain that some of man's own Australopithecine ancestors were using bones as weapons against other animals and fashioning crude flint tools. Their brain size was about 650 to 680 c.c. which is little more than that of the modern gorilla (le Gros Clark[53]).

From this stage to that of the human with a brain of some 1,500 c.c. is an enormous jump and some people argue that men possess abilities which are not even dimly foreshadowed in the behaviour of apes. Nevertheless man's ancestors were at the ape stage relatively recently, and there were successful intermediates like *Homo erectus* whose brain size was 1,000 to 1,200 c.c. Harlow[96] points out that man's use of symbolic language gives him a great advantage over the non-lingual apes, but this gap does not imply that there was anything other than a step-wise progression to the human condition. Perhaps the modern apes are only just below the level for language. Kellogg[135] describes several unsuccessful attempts that have been made to teach chimpanzees to speak. Nobody has succeeded in getting more than one or two recognizable words out of them. But this is probably due to their inability to articulate properly (the human larynx and buccal cavity are considerably specialized for speech, see Lennenberg[149A]) and because communication by gesture is much more natural for them. Recently, in fact, two remarkable long term educational exercises with individual chimpanzees have succeeded in proving that they can use a symbolic language with remarkable facility. The Gardners[81A] have used the language of gesture developed for the use of deaf people; Premack[207A] has used coloured shapes to represent objects and words. In both cases the experimenters have concentrated on training one female chimp, not just to recognize objects and associate them with gestures or coloured-shape symbols, but also to use these same gestures or symbols to ask for things and communicate with the human experimenter. Both chimps have acquired a useful 'vocabulary' of some dozens of 'words'.

These important experiments offer fascinating future possibilities, although they require a formidable investment of time and patience. Suppose one taught the same language to two chimpanzees and left them alone together—what might they 'say' to each other? As it is the results already obtained have forced us to revise completely our estimates of what a 600 c.c. brain can achieve.

9

Learning—2

The most remarkable property of the nervous system is its ability to store the effects of experience. In the last chapter we have discussed the situations in which animals learn and something of what they learn. Here we shall consider the mechanisms of learning and memory storage. Although we shall be using some physiological concepts, we are still far from a complete explanation in physiological terms. Much of the most crucial evidence is still obtained by observing the behaviour of animals in experimental situations, and for this reason, if for no other, the topic is directly relevant to the study of behaviour. In fact, the nature of learning and memory is one of the most important outstanding problems of biology and it is beginning to attract an increasing amount of attention. John[131D] provides a comprehensive survey of this whole field.

As we have been discussing in the previous chapter, a wide variety of animals with very different types of nervous system can benefit by their experience and store memories. The collection of papers edited by Thorpe and Davenport[256] describes a number of experiments on learning with lower invertebrates and reveals the controversy that surrounds their interpretation. Some of this controversy undoubtedly arises because we usually think of learning very much in terms of vertebrates. Natural selection has no respect for methods, it has concern only for end results and we should not necessarily expect the mechanism of long-term adaptive changes as the result of experience to be the same in a flatworm, with its diffuse, non-centralized nervous system, as in an octopus or a rat. Perhaps then we should not say that flatworms *learn*, but rather than quibble over terms it is best to keep an open mind and look out for any useful analogies or similarities between different groups. Almost all the evidence we have from higher animals, where learning takes the familiar form outlined in the last chapter, comes from a single invertebrate—the octopus—and various mammals, including man.

We shall attempt to divide the discussion under two headings, one concerning the neural mechanisms involved in the learning process and the recall of memories, the other on the physical nature of the memory store itself. However, these two topics are closely interrelated and many pieces of evidence are relevant to both.

THE NEURAL MECHANISMS OF LEARNING

When a young game-bird crouches on the first exposure to the sound of its parent's alarm call, it must utilize pre-set, inherited pathways in its brain whereby a particular auditory input has easy 'access' to the motor system controlling crouching. When a rat learns to press the bar of a Skinner box new pathways must become established, because prior to learning the bar evoked no special response. We know, too, that mere presentation of the bar to the rat is not enough, there must be reinforcement. Some part of the learning mechanism records the result of the rat's reaction and if this result is 'good' it increases the probability that the reaction will occur again the next time the situation is presented. Further, we know that somewhere in the nervous system there is stored a more or less permanent record of the learning that can be 'consulted' or recalled on future occasions. This record is usually called the 'memory trace' or 'engram'. Following Young[289] we may collectively call the mechanisms behind learning, storing and recalling events a 'memory system'.

The cerebral cortex of mammals

Because, as outlined in the last chapter, the extent of a mammal's learning ability is roughly proportional to the extent of its cerebral hemispheres, there can be little doubt that these are involved in the memory system. Perhaps a mammal without them can learn—there have been a number of claims that the spinal cord can be conditioned, for example—but such behaviour has little relevance to normal organization. Since most of the evidence to be discussed in this section is based on mammalian work, a brief description of the structure of the cerebral hemispheres is necessary.

They develop as a pair of hollow vesicles from the roof of the fore-brain (see Fig. 4.6, p. 86) and in higher mammals are so large that they come to cover the entire brain. The walls of the hemispheres—their cortex—are thin, only a few millimetres thick and their area is greatly increased by the convolutions of their surface. These form deep grooves or sulci over the

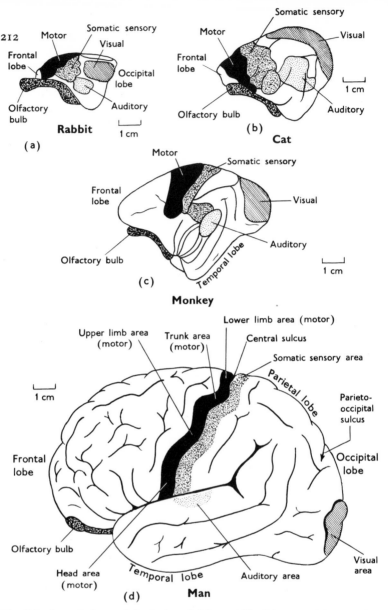

Fig. 9.1 Lateral views of the brains of (a) a rabbit, (b) a cat, (c) a monkey, and (d) man, showing the increasing development of the cerebral hemispheres and the relative reduction in the proportion of their surface devoted to simple motor and sensory control. (a–c modified from Rose, J. E. and Woolsey, C. N., 1949, *Electroenceph. clin. Neurophysiol.*, **1**, 391 ; d modified from *Cunningham's Manual of Practical Anatomy*, vol. III, 11th edn. Oxford University Press, London.)

surface of the hemispheres (see Fig. 9.1). The cortex receives 'projections' of fibres from lower parts of the brain by way of nuclei in the thalamus beneath it, and in turn sends efferent fibres back to the lower centres. The former projections are of two types: firstly, those that come from the sense organs in the form of 'specific projections'; every sensory event perceived is sent up to the cortex, and we can map out 'sensory areas' on its surface (Fig. 9.1) according to what information they handle, auditory cortex, visual cortex and so on. An even more detailed map is possible for example, because it has been found that the visual cortex receives information from the retina of the eye in an orderly fashion; stimulation of adjacent parts of the retina evokes activity in correspondingly adjacent parts of the visual cortex—the latter has a projection of the retina over its surface. Secondly, there are a whole series of 'non-specific' projections which the cortex receives from the reticular formation of the mid-brain and the fore-brain. This reticular formation, as was mentioned in Chapter 3, receives collaterals from all the sensory fibres and serves to 'arouse' the cortex. These non-specific or diffuse projections go not only to sensory areas but also to the so-called 'association areas' of the cortex which lie between them.

The efferent fibres leading from the cortex mostly come from the motor area, which controls all the muscles of the body, and in the same point-for-point fashion as outlined for the visual cortex. A very large fibre tract, the corpus callosum, joins the two hemispheres near to their bases, see Fig. 9.2. Around the corpus callosum, the medial surface of each hemisphere and the base of the fore-brain contribute to the limbic system mentioned on p. 91, which has links with both hypothalamus and reticular formation. Some of it can be seen in Fig. 9.2; Walsh[272] has more details and diagrams.

As can be seen from Fig. 9.1, the more advanced mammals have a decreasing proportion of their cerebral cortex concerned with the special senses and control of muscles. The association areas become very large, especially in the primates; these areas receive many horizontal fibres from other parts of the cortex and were so named because they were thought to 'associate' or co-ordinate diverse sensory information.

The search for the engram

Historically the investigation of the mechanism of learning has centred around the search for newly-made connections in the brain and for the storage site of memory traces or engrams. Pavlov, for example, originally thought in terms of direct connections formed between the sensory 'centre' for the conditioned stimulus in the cerebral cortex and the 'centre' for the unconditioned stimulus. Once this connection was forged, the conditioned stimulus would have gained 'access' to the lower centres, which were already connected to the unconditioned stimulus centre and thus controlled the motor patterns of the unconditioned response. This theory was shown to be inadequate when multiple cuts through the depth of the cortex, made so as to break any connections of this direct type, proved to have little effect either on the acquisition of new responses or the retention of responses previously learnt. Experiments of this type[145] focus attention on a feature of the vertebrate nervous system we have already noted in Chapter 4 (p. 94), namely its extraordinary ability to compensate for the

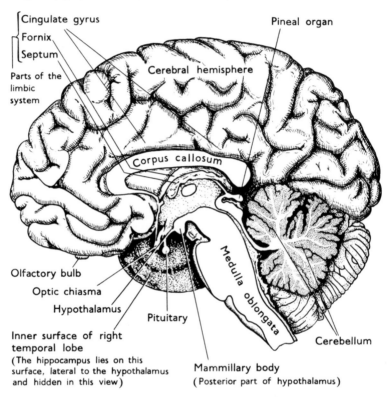

Fig. 9.2 Saggital section through the human brain. The corpus callosum and some parts of the limbic system can be seen. (Modified from Romer,[216] 1962, *The Vertebrate Body*, 3rd edn. W. B. Saunders, Philadelphia and London.)

effects of damage. If a part of a monkey's motor cortex controlling hand movements is removed, the hand of the opposite side is at first paralysed. However, control slowly returns and in many cases a few weeks after the operation the monkey's hand movements are indistinguishable from normal. If the cortex is the main site for the storage of engrams could they show the same resilience to damage? If they did, this in itself would tell us something about their nature, because we should have to abandon the idea that a specific engram depended on specific connections.

The first systematic experiments on the effects of brain lesions on learning seemed to confirm this contention. For a number of years K. S. Lashley,[145] working with rats, investigated the retention of maze learning after parts of the cortex had been removed. One of his experiments used a maze with

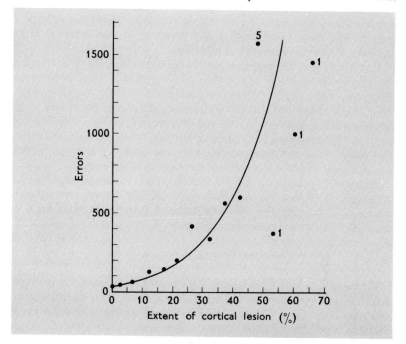

Fig. 9.3 The relationship between errors made during relearning a maze and the extent of damage to the cerebral hemispheres in the rat. Before their operations all the rats had learned the maze well. For lesions above 45% the number of cases (indicated by numerals on the graph) is too small for reliability. All the other points are averaged from many cases. (From Lashley and Wiley,[146] *J. Comp. Neur.*, 57, 31, Fig. 2.)

eight blind alleys on the way to the goal box. All rats were trained to their maximum performance, and then a part of their cerebral cortex was removed—equally and symmetrically from both hemispheres. When the rats had recovered from the operation they were retested in the maze and their errors scored over a number of re-learning trials. Figure 9.3 shows that lesions amounting to some 10% of the cortex have little effect, but beyond this errors begin to rise and above 40% one could say that very little of the rat's original learning remained. The curve is even steeper for more difficult mazes.

The surprising feature of Lashley's results was that the extent of the deficit in memory depended on how much cortex was removed and not what part was removed. Clearly the 'engram for the maze' was not located in any one area. But maze learning depends on a rat using a variety of cues —visual, tacile, olfactory, kinaesthetic—and large areas of the rat's cortex are concerned with handling sensory information. Might not defects in running the maze simply reflect how much a rat was deprived of its senses? Lashley discounts this explanation.

He claimed that blinding a rat peripherally by cutting the optic tract has less effect on its ability to learn a maze than removing its visual cortex, even though we know that the latter operation does not make a rat completely blind. Lashley believed that the visual cortex did not simply handle visual information but, together with the rest of the cortex, facilitated the learning process. He considered that, as far as learning was concerned, the cortex was *equipotential* and that engrams were not stored at any one cortical site but were in some way present throughout its substance.

Zangwill[291] provides a critical discussion of all these experiments. Although they have proved inadequate in some ways, nevertheless Lashley's work forms something of a landmark in the study of the brain. It finally put paid to old comparisons between the brain and a telephone exchange with point-to-point connections. It forced people to think in terms of cortical neurons working in groups or masses for which, indeed, a knowledge of its structure should have prepared them.

Certainly data from humans and some animal experiments using simple tasks involving predominantly one sensory modality, show that equipotentiality is not complete. For instance, it has long been known that damage to the left cerebral hemisphere in right-handed persons has more severe effects on memory and intelligence than equivalent damages to the right hemisphere. This is particularly striking when the areas controlling speech are affected and the division of function between the two hemispheres is now the subject of intensive study, (see Sperry[250A] and Gazzaniga[81B]).

The 'split-brain' technique

Recently a more detailed investigation of the way the cerebral hemispheres function in memory has become possible using a technique developed by Sperry[250] and his co-workers. This involves splitting the cerebral hemispheres apart by a cut down the midline, which severes the corpus callosum, the large band of fibres which links them ventrally (see Figs. 9.2 and 9.4). The split can be extended back through the brain to include even the cerebellum without any serious result. The corpus callosum is the largest single fibre tract in the brain of mammals and it used to be puzzling that complete severance of this tract in man did not produce any detectable changes in behaviour. The reason for this anomaly is now clear because it requires a special type of situation, never normally encountered, to reveal the function of the corpus callosum.

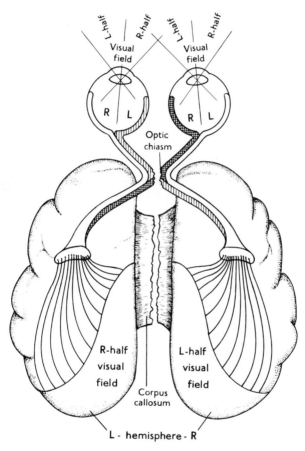

Fig. 9.4 Split-brain preparation seen in dorsal view. Normally fibres from the median half of each retina cross over at the optic chiasma and go to the visual cortex of the opposite hemisphere. Cutting the chiasma restricts the visual information entering each hemisphere to that from the external half of the retina on its own side. When the corpus callosum is also cut, no visual information at all can pass from one side of the brain to the other. (Modified from Sperry,[250] 1961, *Science*, 133, 1749–57. © by the Amer. Assoc. for the Advancement of Science.)

If we learn a visual discrimination with one eye covered, we have no trouble in making the same discrimination subsequently with the other 'untrained' eye. Fibres from the left and right halves of the retina of each eye cross over as shown in Fig. 9.4 and so the visual cortex of each hemisphere receives information from both eyes. If the optic chiasma is split, things learnt with the left eye are still 'known' using the right so long as the corpus callosum is intact. If this too is severed, as in Sperry's 'split-brain' animals, then one-half of the brain no longer 'knows' anything learnt by the other half subsequent to the operation. 'Split-brain' animals are fascinating material because they effectively function as animals with two completely separate brains. It is possible to train the two halves of the brain to make opposite discriminations. Thus the left half may learn that circles are rewarded and crosses not, the right can be made to learn the reverse. It is even possible, using highly ingenious apparatus, to train the two halves simultaneously without their interfering with each other. If forced to choose between its two half-brains, a monkey shows little sign of muddle; one hemisphere or the other 'takes over' at least for a time and the monkey chooses consistently.

The relevance of this work to memory is twofold. Firstly, it shows that one function of the corpus callosum is to transfer engrams from one hemisphere to the other and that storage is eventually bilateral. Normally, of course, information will flow in both directions. The two cerebral hemispheres may still be linked by fibre tracts at lower levels of the brain, but apparently only relatively simple information can be passed across by this route (see below). Secondly, the split-brain technique enables us to explore the rôle of isolated pieces of cortex in memory. If one half-brain is left completely intact it can serve to keep the animal alive and functioning well, whilst the other half can be subjected to much larger lesions than are normally possible. Unless the brain is split, lesions made on one side only are useless because the other side takes over the whole function. With the split brain it is possible to leave a small 'island' of isolated cortex and see what it can store, rather than cut a piece out and see what is missing. Using this technique it has been possible to show that different parts of the cortex are not equipotential. An isolated island of frontal cortex, together with parts of the sensory and motor areas, can store the engrams for touch discrimination in the cat. The whole isolated visual cortex cannot store the engrams of any visual discriminations save of the simplest light versus dark type.

Sechzer[236A] has used split-brain cats to investigate how speed of learning and retention are affected by having only one hemisphere in use. He makes use of the intriguing fact that only with more complex discrimination is transfer of information between the hemispheres abolished by section of the callosum. Split-brain cats learning a simple brightness discrimination

with one eye can transfer to the other perfectly adequately—apparently
the information can be passed to the opposite hemisphere via the lower
parts of the brain—but cannot do so if they are learning to discriminate
between two patterns. In all cases, the cats were trained using one eye to
a criterion of 9 correct trials out of 10, for 3 successive days of testing.
Sechzer found that for pattern discriminations where there was no transfer,
the cats took far longer to learn than unoperated controls (also learning with
one eye) and they were also poorer at retaining what they had learnt. When
learning a brightness discrimination, which does get transferred, the split-
brain cats learnt at the same speed as normal animals (and retained just as
well). This shows that there is nothing about the operation itself that could
account for their slowness in the non-transfer situation. These results
certainly suggest that having access to twice as much neural tissue helps
both in acquiring an engram and in storing it, although as we have seen
different parts of the hemispheres apparently vary in their storage
capabilities.

Quite soon it should be possible to answer a whole series of questions on
the memory system using split-brain animals. How long does it take to pass
an engram through the corpus callosum? If some of its fibres are cut does
the engram become imperfect, or merely take longer to get through? What
connections are necessary between cortical islands and sub-cortical
structures if the former are to store engrams? Answers to such questions
will tell us a good deal about the nature of engrams and the memory system
generally.

Long-term and short-term memories

One of the most striking features of the mammalian memory system is
that memories of recent events have some different properties from
memories of events further in the past. The former are much more susceptible to disturbance of various kinds. People suffering from concussion are
often unable to recall the immediate events which led up to their accident,
although they can recall events in the past quite normally. If their recent
memory recovers, and it does so in many cases, memories return
roughly in order—the most distant ones first—until gradually all the events
before the accident can be recalled. This phenomenon is called 'retrograde
amnesia' and it can be reproduced in animals.

For example, one common technique has been to subject rats to a brief
electro-convulsive shock (ECS) given by passing a current through the
brain between electrodes clipped on to their ears. The ECS is given following a learning trial and its effects on memory can be tested by comparing
the performance of shocked rats with control animals which merely had
electrodes clipped on. Testing 24 hours later, when the shocked animals

had fully recovered from the effects of their convulsion, it has commonly been found that their performance is grossly impaired and this might suggest that they do not retain any memory of the previous learning trials.

Phenomena of this type have led many workers to support a 'consolidation hypothesis' which proposes that events are recorded in two ways. First they enter a short-term memory system which persists for up to an hour or more but then decays. Secondly, whilst the events are still accessible to recall from the short-term memory, they are also undergoing transformation or consolidation into a long term engram. This engram may be formed from the short term memory's record itself, or perhaps from a parallel record. In any event, according to the hypothesis during the consolidation period which may take a matter of minutes or hours, the engram is labile, sensitive to disturbance and perhaps destruction. One common supposition has been that short-term memory and consolidation itself relies on continuous activity in neurons, but the long term engram, once stored, does not require such activity and is consequently less easily disturbed.

In trying to test this hypothesis, we have from the outset to face the awkward fact, alluded to already on p. 176, that it is extremely difficult to distinguish between poor consolidation leading to a weak or incomplete engram and poor recall of a normal engram. Although experimenters may have assumed they are studying storage, all the experimental treatments that affect memory could be acting on recall. If so it remains true that recall for recent events appears easily disturbed by treatments which leave recall of events long past unaffected.

The consolidation hypothesis, in its various versions has engendered a large amount of research. Unfortunately it cannot be said to have reached any incontrovertible conclusions and at present there is disagreement even about some of the basic phenomena. Thus ECS experiments of the type outlined above have been widely criticized on a number of grounds. For one thing there can be no doubt that rats find ECS extremely unpleasant. Hence it is possible to confuse their fear at being in a situation where they received an ECS previously with loss of memory for prior events.

To get round this difficulty, the 'step-down' or 'step-out' experiment has been used in conjunction with ECS. Thus Madsen and McGaugh[170] put rats on to a small elevated platform above a grid. After a short time the rats stepped off and when they did so they closed a circuit and their feet received a painful electric shock. A normal animal placed on the platform a second time does not step off again, he retains the memory of the last trial. Madsen and McGaugh gave half their rats on ECS immediately after they had stepped off on the first occasion and received a shock. If the results of an ECS were to induce fear, this should summate with the effects of the shock to the feet and make such rats even less likely to step off a second time. In fact rats given ECS step off without hesitation. This certainly looks convincing evidence for ECS abolishing memory, but there are other possibilities. It has been suggested that one effect of ECS is to increase locomotor activity in rats, so that shocked animals are likely to step down from a platform more quickly than controls even if they remember just as well. Zornetzer and McGaugh[293] refute this explanation for the effects of ECS in mice and provide a useful summary of other recent work on both sides of the dispute.

A more serious criticism of ECS experiments concerns the permanency of their effects. Several workers (see references in Deutsch[66]) have reported that memory for past events, apparently banished by ECS, can under some conditions re-appear later. Now the simplest form of consolidation hypothesis would predict that once destroyed by ECS (as revealed by loss of memory just afterwards) there should be no trace left upon which any subsequent consolidation process could build. There is a direct conflict of evidence here because McGaugh's group find that amnesia is permanent under their experimental conditions.[60E] We must simply wait and see. Of recent studies, that of Quartermain et al.[207B] is impressive for the range of control experiments they run in their tests on the effects of ECS using the standard avoidance of foot shock situation. They have clear evidence that rats can recover some memory after ECS, if they are given a brief 'reminder' trial without reinforcement in the original situation, followed by a 'reminder' foot shock (given in a completely different situation). Recovery of memory depends on the gap between reminder trial and reminder shock being not more than about 4 hours. There would be little chance of the rat learning from scratch given a trial in one place followed hours later by a footshock given elsewhere. The fact that it does remember on subsequent tests certainly suggests that the reminders operated on some of the original trace that remained intact even after ECS.

McGaugh's group have also approached the consolidation phenomenon in another way: they have experimented with drugs which do not disrupt memory but actually improve it.[165-7] They have been able to *increase* the effects of a learning trial by immediate post-trial injections of strychnine and related substances. At high concentrations strychnine causes convulsions, but a low dose simply increases excitability at synapses in the central nervous system. Their interpretation of such results is that information entering a strychnine-injected rat's brain as a result of a learning trial produces changes which persist for longer than normal. Thus a single trial can consolidate the engram to a greater extent and is equivalent to repeated trials in a normal animal.

Again one might suggest as an alternative explanation, that strychnine improves recall or performance on tests for retention because the rat is still highly aroused by the drug. If this were so then strychnine should be more effective the closer it is given to the retention test. But this is not the case, its effectiveness is greatest when given close to the original learning trials and therefore furthest from the retention test some 24 hours later. This suggests that it is affecting the conditions under which consolidation takes place, beginning immediately after a learning trial. In general it has been found that disruptive treatments like ECS or anaesthesia are also most effective when given immediately after the learning trials. However there are still huge discrepancies in estimates of how long consolidation takes to be complete: they range from a few seconds to a day or more!

A promising line of attack on this problem is that of Cherkin.[49A] He induces amnesia in chicks using carefully graded doses of a convulsant chemical after learning. His evidence suggests that the production of amnesia is not an all-or-nothing matter, but that low levels of amnesic treatment slow down consolidation without stopping it. This, Cherkin suggests, could account for the wide discrepancies in estimates of the time for which the memory trace remains labile and vulnerable to disruption.

The nature of the mammalian memory system

There is good evidence that the cortex is the site of long-term engram storage in mammals, however vague are the details. Does consolidation also take place in the cortex, or are engrams laid down elsewhere and then passed up to the cortex for storage as they are passed through the corpus callosum from one hemisphere to the other? This latter seems unlikely, but quite certainly damage to particular parts of the brain outside the cortex can permanently affect the consolidation of new engrams without affecting those already in storage.

The *hippocampus* (so named because its coiled shape in the human brain recalls that of the sea-horse, *Hippocampus*) lies on the lower inside margin of each cerebral hemisphere and forms part of the limbic system (Fig. 9.2, p. 214). It has direct connections to the hypothalamus and to the cortex via the reticular formation of the fore-brain. Humans with brain damage affecting both hippocampi show severe defects in their short-term memory. The effect is often most striking, as in a case described by Victor *et al.*[269]:

'The patient's ability to recall events and to learn and retain new facts were both seriously affected. He would ask the same question over and over again. He spent hours watching baseball on television, but as soon as the set was turned off, he was unable to remember the score or any other detail of the game. However, he was able to recall correctly the highlights of games that had been played many years before.'

Clearly patients like the above have not lost all ability to recall recent events. If they had they would be unable to talk properly, because by the time they spoke words at the end of a sentence they would have forgotten words spoken at its beginning, and so on. If presented with a series of numbers to remember, such a patient can go on recalling them so long as his attention is rigidly fixed on the task. If he is distracted, even for a moment, the numbers are lost.

It would seem that *attention* is very important and that the hippocampus may be concerned in maintaining it. This conclusion gains some support from the broad connections between the hippocampus and the reticular formation, whose function in arousing the cortex and focusing attention has already been mentioned in Chapter 3, p. 38. In fact lesions in the reticular formation of the fore-brain also affect the memorizing of recent events.

It is possible to suggest an outline hypothesis of the following type. Events in the external world are recorded via the sense organs and information passes up to the relevant part of the cortex and simultaneously to the reticular formation and limbic system. If the recorded events have significance for the animal, their neural effects are maintained by the activity of the reticular formation, and the consolidation of the engram begins in the cortex. The limbic system 'decides' whether the events have significance by recording their 'results'—whether they are associated with a reduction in thirst, whether they are associated with pain, or with sexual satisfaction, and so on. The limbic system is directly connected to the hypothalamus which, as we discussed in Chapter 4, has a regulatory function for all 'bodily needs', and the activity of the limbic system modifies activity in the hypothalamus. We have mentioned on p. 189 that rats and other animals will learn to press a lever in order to receive stimulation at various sites in the hypothalamus and limbic system, other sites in the same systems produce aversion.[202] Here we may be dealing with channels that normally record the 'good' or 'bad' results of an event. According to these results so the cortical engram will form connections with what may be broadly labelled 'approach' and 'avoid' motor channels, which will determine what the animal does when next the same events occur in the external world.

This hypothesis is highly speculative and may infuriate those who dislike the use of phrases such as 'form connections with' in the previous sentence, to which we can give no precise physiological meaning as yet. Nevertheless research into memory systems must proceed at all levels and, although it may be a long time before we can give an explanation in terms that would satisfy a neurophysiologist, it is worth while trying to suggest the *principles* upon which a memory system might be based. An hypothesis of the type outlined above does make use of the anatomic relationships we can see in the mammalian brain, which can scarcely be fortuitous, and serves to stimulate research.

Sometimes we can observe that the neurons of parts of the brain are arranged in orderly and repetitive arrays. This suggests that such order may be involved in the classification and processing of events which are subsequently memorized. Recently several elegant models of memory systems have been constructed which relate very closely to observed structure. That proposed by Blomfield and Marr[32A] which accounts for the operation of the cerebellum in organizing the learning of new voluntary movements is an example of this type. Young, in his work on the memory system of the octopus to be discussed later, has also strongly advocated basing memory models upon observed neuroanatomy.

How much is learnt? The problem of 'recall'

In the foregoing section we have drawn together information gathered from work on a number of different mammals including man. Obviously it is a grotesque oversimplification to interpret the results of learning in ourselves in terms of 'approach' and 'avoidance' tendencies or to suggest that 'punishment' or 'reward' have any simple meaning. The range of possible motivation and rewards for learning are far wider.

It is quite possible that all events which 'engage our attention' at the time are stored, although only a fraction of them are ever recalled. We all know that certain trivial events in our past can be recalled although they certainly were not associated with strong emotional arousal. Presumably they attracted sufficient attention at the time and for some reason, as yet quite mysterious, recall has remained easy. Other events, which may have been equally significant, are apparently inaccessible to recall, if they were ever stored.

It is difficult to investigate the problem of how much of their experience animals store, because we have such limited ways of testing their memories. Humans can tell us what they recall, and the most remarkable results are obtained from directly stimulating the temporal lobes of the cortex in fully conscious patients. Penfield[205] records how, when stimulated, a person would report 'hearing' a tune or 'seeing' some event in the past as it happened. These were not just brief effects—the music persisted and could be elicited again when stimulating the same spot on the brain later. The patients claimed to identify these impressions as hitherto unrecalled memories of actual events precisely as they occurred. This is possible, and if true would indicate that our brains store much that is inaccessible until recall is set in action by artificial stimulation of the cortex. However, the details of the impressions could be partly hallucinations based on scraps of recalled material; there was no means of checking. Hunter[123] gives examples of the fictional embroidery with which, quite unintentionally, our minds clothe partly recalled events.

In the last chapter we discussed two ways to test 'remembering' in humans—'recognition' which involves testing whether a certain situation is familiar and 'recall' which term, as used by psychologists, means that a subject spontaneously reproduces in writing or in speech what he remembers. Hitherto in this chapter we have used the term 'recall' to include both types of remembering and indeed it seems most unlikely that they are fundamentally different. In both cases recognition must be involved, although it need not be recognition of events in the outside world. In some way our brain can 'reproduce' sensory events within itself and these interact with engrams where they 'fit' and evoke recall. At present the nature of the recall mechanism is still quite obscure. Recall is possible within a few seconds of an event occurring, at a time when we know the engram is susceptible to disturbance and sometimes destruction. It can operate within a few milliseconds when working well or it may temporarily fail us altogether, only to work later without any conscious effort on our part. Someone's name is 'on the tip of our tongue' but eludes us, only to come back suddenly when we are thinking about something else.

The memory system of the octopus

The outline hypothesis for the mammalian memory system given earlier owes much to J. Z. Young[288-9] who bases it by analogy from extensive work on the octopus by himself and a number of other workers.

Both an octopus and a rat will learn to approach one set of objects which are associated with food and avoid others which are associated with electric shock. They both have a brain made up of neurons whose structure and physiology are very similar. It is therefore entirely appropriate to see whether they solve the same problems in a similar fashion. If, as is the case, there are similarities in the arrangement of functions within the two brains, then it is reasonable to suggest that they may operate on the same principles. At the level of the sense organs, the eyes of octopuses and mammals show astonishing similarities which are entirely the result of convergent evolution. The study of comparative anatomy and physiology is full of similar examples with totally unrelated animals responding in the same way to the same demands of their environment. We can profitably look for convergent evolution in brain and behaviour.

The cephalopod molluscs, to which group the octopus belongs, have large brains. They are mostly active, predatory animals which hunt by sight, although the octopus has in addition a very extensive tactile system with its prehensile arms.

The brain of the octopus is illustrated in Fig. 9.5. Its main features are (a) the enormous optic lobes which are situated just behind the eyes and are joined to the rest of the brain by broad optic tracts, (b) the sub-oesophageal lobes linked to the nerves of the arms and mantle which control the movements of the animal, and (c) the supra-oesophageal lobes which surmount the others, the vertical lobe being the most dorsal with the frontal lobes beneath it.

Crabs probably form a major part of the octopus's natural diet and it will readily emerge from a shelter at one end of an aquarium tank to attack crabs presented at the other. Using the techniques of conditioned discrimination (p. 186) it has been possible to investigate how well an octopus can distinguish shapes displayed alongside a crab as it is lowered into the tank. An octopus will readily learn to avoid 'crab plus horizontal rectangle' if attacks in this situation are always punished with electric shocks. The same animal will continue to attack 'crab plus vertical rectangle'.

If one of the supra-oesophageal lobes—the vertical—is removed from an octopus after it has learned such a discrimination, it attacks whenever a crab is presented whether with a horizontal or a vertical rectangle. Apparently it has 'forgotten' its learning and we might deduce that the vertical lobe was the site of the engram store. However, the position is not so simple as this, because if two crabs are presented to the octopus *simultaneously*, one with a vertical, the other a horizontal rectangle, it usually attacks the 'correct' one. The loss of the vertical lobe makes the animal much more likely to attack, but its engrams are stored elsewhere.

If an octopus without its vertical lobe is shocked for attacking a crab with a horizontal rectangle, it retreats and will not attack again for a time, but the effect has worn off after 5 minutes and it will then attack again without hesitation. This process can be repeated indefinitely; given trials once every 5 minutes or longer the octopus shows no sign of learning. However, given trials every 2 minutes it learns slowly and once it has learnt, retains its memory quite well. It still tends to attack too readily, but particularly when tested with simultaneous presentation of 'right' and 'wrong' objects it responds correctly.

Further experiments show conclusively that the engrams of visual learning are situated in the optic lobes themselves, and a short-term memory is set up there after each presentation of stimulus and punishment, but this fades after a few minutes unless the vertical lobe is intact. Rapidly repeated trials can substitute for the action of the vertical lobe to some extent.

There are large fibre tracts leading from the optic lobes up to the supra-oesophageal lobes, and from there further tracts go back to the optic lobes again. This arrangement suggests that the optic lobes and vertical lobe may have a functional relationship rather like that of the cortex and the limbic system in mammals. The vertical lobe serves to 'maintain attention' on the signals resulting from each trial and thus facilitates the consolidation of the engram in the optic lobes. Within these lobes the engram, whatever its nature, is fairly evenly distributed. Lesions have effects on retention

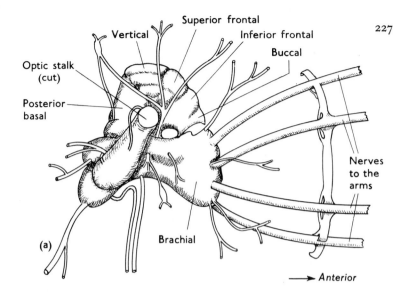

Vertical

Superior frontal

Inferior frontal

Buccal

Optic stalk
(cut)

Posterior
basal

Nerves
to the
arms

(a)

Brachial

→ Anterior

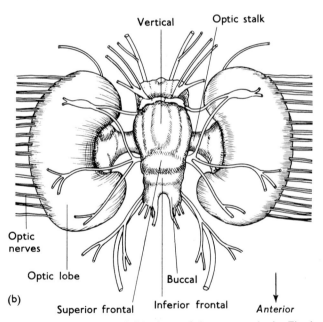

Vertical

Optic stalk

Optic
nerves

Optic lobe

Buccal

(b)

Superior frontal

Inferior frontal

Anterior

Fig. 9.5 Lateral (a) and dorsal (b) views of the octopus brain. The brain surrounds the oesophagus which runs through it just below the buccal ganglion and
optic stalk. Note the enormous optic lobes which contain well over half the
number of neurons in the whole brain. (From Young,[288] 1964, *A Model of the
Brain*. Oxford University Press, London.)

roughly proportional to their size and not their site, rather reminiscent of Lashley's early experiments with the mammalian cortex.*

The analogy with the mammalian system can be pressed further because the vertical lobe, like the limbic system, also seems to be involved in the recording of the results of action—in this case whether a shock or a food reward follows the presentation of the stimulus and subsequent attack by the octopus. Fibres which probably record pain and food sensations branch and enter both optic and supra-oesophageal lobes. We have seen that removal of the vertical lobe increases the likelihood of attack. Below it lies the median superior frontal lobe, damage to which reduces the attack tendency. In the normal animal these two opposing centres will perhaps 'strike a balance' between attack and retreat depending on incoming signals of the results of an action.

It is very difficult to record the activity of neurons in the octopus because it has no adequate skeleton for anchoring electrodes. Hence most evidence on the function of fibre tracts is based on detailed histology and careful work with lesions. There are still many problems but, because much of the octopus's behaviour can be analysed in terms of 'approach' and 'avoid' tendencies, we are certainly closer to understanding the principles on which its brain 'works' than in any other animal. A full appreciation of this important research can be made from excellent reviews by Young[288-9] which give complete references.

THE PHYSICAL NATURE OF THE ENGRAM

Many of the experiments we have been discussing tell us a good deal about what the engram is not, without giving much positive information. If the engram has 'units' these are likely to be diffused through a considerable volume of nerve tissue although not through the whole brain. Numerous theories have been suggested and here we have space to deal with only a few.

It is still attractive to think in terms of new connections being built up within the brain, even though they must be built at numerous sites rather than singly. Young[289] has found numerous large neurons with two axons in the optic lobes of the octopus, which are very rare elsewhere in its brain. He suggests that these are units of the visual system which react to a particular stimulus and have alternative outlets to 'approach' and 'avoid' systems. During engram storage, one of these outlets is permanently inhibited. When enough of these memory units, or 'mnemons' as he calls them, have

* Wells[281A] describes some experiments using the split-brain technique on octopuses. As in mammals, engrams appear to be stored bilaterally and, at least for tactile learning, it takes about an hour for an engram to be passed across the midline.

been 'switched' the whole animal's behaviour will be biased in one or other direction, in response to particular visual stimuli.

Whatever else may be uncertain about the engram there can be little doubt that we are looking for a permanent growth process; nothing else can account for the extraordinary stability of engrams. Once established they can last for 70 years or more and withstand a variety of brain damage.

Examination of the detailed structure of brain areas known to be sites of storage for engrams such as mammalian cerebral cortex[243] or octopus optic lobes,[289] reveals numerous 'self-exciting' or 'reverberating' circuits; i.e. neurons arranged in a loop so that impulses transmitted along the axon of a neuron eventually get conducted back to its cell body or dendrites, thus— potentially at least—the cycle begins again. At one time it was suggested that the engram was represented by continuous activity within such circuits. Whilst it is very probable that they are involved in the consolidation process, it is very unlikely that they can be used for storage over long periods. Nerve cells are efficient transmitters within their brief millisecond time scale, but accurate repetition of a highly specific pattern of impulses would break down after minutes, let alone hours or years. Perhaps the neatest demonstration that continuous electrical activity in the brain is not necessary for maintaining engrams comes from work by Andjus *et al.*[7] They cooled rats down to O°C for periods of up to an hour, at which temperature all electrical activity in the nervous system ceases. Nevertheless, on recovery these animals retained their memory of events prior to cooling as well as normal rats.

If continuous activity is ruled out, it is still possible that electrical activity associated with a learning trial 'wears a pathway' in the brain which will facilitate transmission along this route the next time the events recur. There is good evidence, from peripheral nerves at least, that activity does facilitate transmission. Axons and nerve endings swell to a small but significant extent and this increases the area of contact between one neuron and the next.[69] However, all such changes are short-lived and persist only for a few hours unless continuous activity is maintained. Again one might suppose that these short-term facilitatory effects are involved in the consolidation process but since they are reversible, they are not an adequate basis for long-term storage.

Protein synthesis during the formation of engrams

If a permanent growth process is involved when an engram is laid down, it seems unlikely that it is reflected in large structural changes although there may be some growth of new synaptic connections between neurons. More probably, some connections which are already there become facilitated, and this will involve permanent chemical changes within the neuron so that transmission across particular synapses is made easier. In the nervous system chemical changes are in some ways easier to detect than structural ones, and exploration along these lines which has led to some advances in the study of the engram. Most work has concentrated on the rôle which substances involved in the synthesis of new proteins might play in laying down the engram. Living cells are continuously breaking down some of their proteins and building new ones, and neurons are no exception. Transmission across synapses involves the synthesis of a 'transmitter substance'—known to be acetylcholine in many cases— which diffuses rapidly through the synaptic membrane and can then initiate an impulse in the adjacent neuron.[69] Both the synthesis of acetyl- choline and its rapid breakdown once it has crossed the synaptic mem- brane, require enzymes which are proteins and thus there must be quite extensive protein synthesis. We therefore expect active neurons to be synthesizing protein, but evidence suggests that during the consolidation of engrams the rate is higher than for the normal needs of transmission. Something else must be going on—perhaps the synthesis of new transmitter substances or changes to the synaptic membranes, either of which might increase the efficiency of transmission.

The 'biochemistry of learning' is now attracting a great deal of attention. but unfortunately much of the work is open to all the doubts and diffi- culties which we have already discussed in relation to ECS and other forms of disruptive treatment on short-term memory. The key reference here is Glassman[82B] who provides a most admirable, balanced review of the whole field.

In fact, the experimental procedures used to investigate the rôle of protein synthesis in memory storage are very similar to those used for physical disruption of consolidation. For example the drug puromycin, which specifically inhibits protein synthesis, also interferes with memory. Mice are taught that to avoid an electric shock they must run into the left arm of a Y-shaped maze. They will learn this task rapidly; most animals reach a criterion of 9 correct responses out of 10 within an hour. At varying periods after learning they are given an injection of puromycin directly into the brain (controls receive saline) and their performance is tested later. Retention is usually scored by comparing the number of trials they require in the Y maze in order to reach the 9 out of 10 criterion again. Con- trol injected mice retain their memory very well but those which received puromycin just after the learning trials show little sign of memory a few hours later.

The puromycin thus has effects which resemble superficially at any rate those of ECS or anaesthesia, especially as old-established long-term memories are not affected by the drug. Does puromycin (only one of several drugs with comparable effects) affect consolidation of the engram, with the implication that this involves synthesis of special proteins? It may do, but it may produce its effect by disrupting the recall system or the learning process itself. As mentioned above, the mice have their post-treatment performance tested in the Y maze and memory deficit is measured by the number of trials to reach criterion level again. This means that if puromycin still in the brain affected their learning, this would be confused with loss of previous memory. Earlier work[73] suggested that the drug did not affect learning, but this is no longer certain. Nor is it clear how permanent is the amnesia induced by drugs. Here, as with ECS experiments, there is now evidence that something of the engram survives the blockage of protein synthesis.

We should not be too depressed by this temporary uncertainty; the picture will clear as research proceeds. It is simply that in this field it is very difficult to design experiments that yield watertight results. In spite of the uncertain state of current knowledge, it would seem highly probable that the synthesis of new proteins *is* involved in engram formation, leading to structural changes in membranes or changed rates of enzyme production which will alter synaptic transmission.

RNA and memory

The past decade has seen tremendous advances in our knowledge of the coding of genetic information and its translation into the synthesis of particular proteins which largely determine the characteristics of different cells. Paul[203] provides a clear account of modern cell biology. The genetic material in all higher organisms is deoxyribonucleic acid (DNA) whose molecule consists of two long chains of alternating sugar and phosphate, twisted round each other to form a double helix. Linking the two chains, like the rungs of a ladder, are combinations of four bases, and the order of the bases along the chain determines the proteins which will be made. DNA itself does not make proteins from the constituents in the cell nucleus; another molecule of a similar type, ribonucleic acid (RNA) is formed using the DNA as a 'template' so that the RNA molecule has the same sequence of bases as the DNA and therefore the same 'code'. This 'messenger RNA' moves through the nuclear membrane into the cytoplasm where it combines with other forms of RNA to organize the synthesis of proteins from constituents there.

Recently there has been a good deal of speculation and experiment on the possible rôle of RNA in memory storage. Several types of observation have been presented as evidence for this; see Deutsch,[66] and Glassman.[82B]

1. As we have just described, protein synthesis occurs during engram consolidation and we know that RNA controls protein synthesis.
2. The cell bodies of neurons contain large amounts of RNA, more than would perhaps appear necessary for routine maintenance and impulse

transmission. (However, some of the measurements quoted do not take into account the considerable volume of the axon which contains little or no RNA for itself.)

3. The RNA in certain neurons of rats which have just been trained to perform a task, changes in its chemical structure as well as increasing in quantity. Passive stimulation without learning merely increases its quantity.

4. Injecting RNA extracted from the brains of trained animals into the brains of naïve animals has sometimes been claimed to reproduce the effects of training in the latter.

Clearly these different pieces of evidence need not all imply the same thing. 1 and 2 follow naturally if building the engram involves growth and a lot of protein synthesis. In this sense RNA is necessary for learning in the same way that, say, oxygen is necessary for learning.

Evidence under 3 and 4 implies that RNA plays a more important rôle than just enabling a neuron to make proteins; it suggests that RNA may actually 'contain' or 'encode' information on the events which are learnt. This seemed an attractive possibility in the mid 1960s and—as suggested in 4 above—a number of workers tried to extract learnt information in a chemical form and inject it into untrained animals. Some claimed success, but more reported failure and attempts are now much less frequent. Glassman reviews the whole story and Frank et al.[75] provide a recent contribution which indicates clearly some of the confounding factors involved when one injects animals with extracts of crushed-up brain cells. In retrospect this approach seems very naïve, for our knowledge of the properties and rôle of RNA do not suggest even remotely that it could serve as a 'memory molecule'.

Although the biochemical approach to learning has proved illuminating in a number of ways, we must remember that ultimately we need to explain how neurons come to change their functional interconnections as a result of experience. To this problem, biochemistry may provide only subsidiary answers; we need much more to investigate neuroanatomy and physiology.

IO

Social organization

The primary aim of this book has been to present a survey of some topics which illuminate the organization of behaviour within the individual animal. However we have frequently had to discuss aspects of behaviour which involve the interaction between individuals, as in aggressive behaviour and courtship. In fact virtually all animals exist in pairs or in larger groups for at least part of the time. For some species, sociality is the dominating feature of their whole lives and we must now consider the social life of animals in its own right.

The cohesiveness and co-ordination of animal societies are often their most striking feature. It is easy to lose sight of the individual in such groups, thus failing to recognize that the forces which shape its behaviour are the same as those for more solitary species. Natural selection operates upon individuals and the responses of a social animal to the other members of the group will evolve to its own best advantage.

The term 'social organization' refers to populations and not to individuals and defines the nature of the interactions between members of a species. In some instances—the various social insects for example—social organization is fairly rigid and species-specific. In vertebrates, as we shall discuss later, it is a much more dynamic phenomenon, and may vary with changing conditions. Certainly use of the term is not restricted to highly social animals. Tigers which usually live and hunt alone in large territories, avoiding contact with others save for breeding, and honey bees which spend their entire life in a dense colony, both provide examples of social organization, albeit of widely different types.

It is perfectly valid to refer to any interaction between one individual of a species and another as 'social behaviour'. This is the criterion adopted by

Tinbergen[264] and Dimond[67A] in their books but here, whilst reviewing social organizations of diverse types, we shall concentrate most attention on what might be called 'animal societies'. A true society will involve more than a mated pair or a mother and her offspring; it will mean a stable group whose members inter-communicate extensively and bear some relatively permanent social relationship to one another.

Within the animal kingdom there are an enormous variety of groupings, not all of which meet the criteria outlined above. Two contrasted examples of undoubted societies are provided by ants and chimpanzees. The ants that make up a colony live in a common nest which they have constructed together. There is a stable relationship between the queen and her sterile daughters, who may be differentiated into two types—often called workers and soldiers—distinguished both by their structure and their rôle in the life of the colony.

Chimpanzee society is much less closely organized. A group of some 15–20 individuals move around together, sometimes splitting up and reforming with other groups nearby. The rôle of individuals in the group is partly determined by their age and sex, but also by their genealogy. Mothers and their successive offspring tend to stay together for many years, and each group usually contains one or more adult males. Individuals live for 30 to 40 years and associate with the same group for most of this time.

Compared with these true societies the organization within a flock of birds or a school of fish is much less complex, although individuals may stay together for months. When we come down to a swarm of water fleas gathered in some area rich in food, or a mass of *Drosophila* collected on some rotting fruit, then quite clearly we are not justified in using the term 'society'. *Drosophila* and water fleas form *aggregations* because they are attracted to a common food source; they certainly react to one another's presence, for example by spacing themselves out so that they do not touch, but they do not constitute a society.

The advantages of grouping

We have already suggested that the forces of natural selection will shape an individual's responses within a group just as they do for any other aspect of its adaptation to its environment. Consequently we may expect that all animal groups, whether aggregations or true societies, confer an advantage on the members that constitute them.

The work of Allee[4] and his school was of great importance in revealing numerous examples of the way in which an aggregation benefits the individuals which comprise it. Sometimes the advantages relate almost entirely to the physical size of a group. Water fleas cannot survive in alkaline water, but the respiratory products of a large group of them are sometimes sufficiently acid to bring down the alkalinity to acceptable levels. Thus a group can survive where a few individuals could not. *Drosophila* cultures do not do well if they begin with too large a number of eggs—the resulting larvae are undernourished—but they fare equally badly with too few eggs, because there are not enough larvae to break up the medium and make it soft enough for them to feed. It is advantageous for a female to lay her eggs close to those of others because her own offspring will benefit.

Bird flocks and schools of fish exemplify groups which, though lacking some of the attributes of a society, often go far beyond simple aggregations. There may be a much higher degree of interaction between the individuals. Physical factors still count as with emperor penguins which huddle closely together as they stand incubating their eggs during the antarctic winter. Heat is conserved and birds on the outside move more than those in the centre thereby leading to mixing and a reasonable distribution of shelter. One of the most obvious advantages of a cohesive group whose members respond to each other's behaviour is protection against predators. With a number of animals on the alert, the approach of a predator is less likely to go undetected and one alarm call will suffice for all, but in addition the group can take concerted evasive action. Fig. 10.1 illustrates this for a flock of starlings and exactly similar behaviour is shown by some fish which bunch together at the least alarm. Predators rarely attack an individual in a close group and their commonest stratagem is to make swoops towards the group which may cause it to scatter, when the predator singles out an isolated animal.

Colonial nesting birds—gulls and terns for example—may provide formidable opposition to an invading predator such as a fox by their mobbing attacks, even though each is responding more or less individually to defend its own territory. More highly organized societies may have even more positive defences against predation; we have already mentioned the protection against leopards which baboons secure by living in groups—several adult male baboons united against it are more than any leopard will face. Plate 3 shows the protective formation of a herd of musk oxen at the approach of wolves, with the young animals sheltered by the adults.

Protection from predation is only one factor conferring advantage upon groups. A colony of prairie dogs can by their continuous grazing keep a large area free of tall grasses thus allowing the smaller herbaceous plants on which they feed to grow more freely. Foraging in flocks may enable birds to find food more easily both because the group will disturb the habitat more effectively and stir up prey animals, and also because the sight of one bird feeding successfully attracts others to the spot. Large predatory mammals such as lions, hyenas and cape hunting dogs also rely on co-operation between members of a group to hunt their prey. Their stratagems may involve some individuals driving prey towards others hidden in cover, or as

Fig. 10.1   The response of a flock of starlings to the approach of a bird of prey. (From Tinbergen,[258] 1951. *The Study of Instinct*. Oxford University Press, London.)

Plate 3 The defensive formation of a group of musk oxen on the Canadian tundra. When a predator approaches, they bunch with the older animals at the front, facing the threat. (Photo by D. Wilkinson from Information Canada Photo-thèque.)

with the dogs, taking it in turns to run down an antelope to the point of exhaustion.

Another factor which may make group life advantageous concerns the stimulation and synchronization of breeding. In Chapter 6 we discussed experiments which showed that the hormonal cycles of some female birds and mammals are affected by the presence of males. Large numbers of sea-birds nest in dense colonies, even though they may vigorously defend their own small territory within the group. Fraser Darling[60C] was among the first to point out that in such colonies a great deal of stimulation must result from the displays, both visual and auditory, which are a constant feature of a large group early in the breeding season. The effects of such stimulation, coming from neighbours as well as from a bird's own mate, will tend to accelerate and synchronize the reproductive cycle within the colony. This, Darling suggested, was advantageous because it reduced the vulnerable breeding season to a minimum. If it leads to synchrony in egg laying and hatching then predators, who can only take so many prey each day, will be 'swamped' by the sudden brief flush of eggs and chicks so that the chances of an individual nest being attacked are reduced.

Experimental verification of the so-called 'Fraser Darling effect' is not easy to come by, but Coulson's[54D] remarkable long-term studies of kitti-wakes go some of the way. He has shown that birds breeding in the most densely-packed central areas of a nesting colony are more successful than those in occupying more peripheral nest sites. They lay more eggs and lay them earlier and there seems little doubt that heightened social stimulation in the central area is one factor contributing to these effects. The centre exerts great attraction and early in the season kittiwakes compete fiercely for sites in this area.

Finally in our discussion of the advantages of group life, we must briefly consider the views of Wynne-Edwards set out in his important book *Animal Dispersion in Relation to Social Behaviour*.[287] Wynne-Edwards takes a radically different view from that expounded earlier in this chapter. Here we have discussed the benefits which social life confers upon the indivi-duals which make up the group. Wynne-Edwards suggests that this is not the whole story and that commonly during the evolution of social life, the 'selfish' needs of the individual have been sacrificed for the benefit of the group as a whole. Only in this way, he suggests, can animals live within the resources that their environment provides. If individuals always re-produced at the maximum rate they would rapidly eat out their food supply and Wynne-Edwards asserts that this almost never happens in natural circumstances. Groups have evolved in which breeding rate is linked to the available resources. Animals maintain their breeding rate below the maxi-mum and the most important regulatory mechanisms are to be found in their social systems. Because groups are stable and to a large extent reproductively isolated from others, selection has operated on the groups themselves, weeding out those which did not adapt to match their resources.

In Wynne-Edwards' view a good proportion of social behaviour, ie. the communication and interaction between individuals, serves this function of regulation. It may be the familiar type of territorial behaviour in which a male, be it fish, bird or mammal, defends an area against all rivals. Here Wynne-Edwards would claim that the territory acts as a symbolic feature determining an animal's status in the population. Only those which hold territories can breed and, though not related directly to the food supply, territorial behaviour limits the number of breeding individuals in a way which relates ultimately to how many animals the environment can support.

On the other hand he would also claim an analogous function for many of the elaborate communal displays of animals. For example, the extraordinary aerial manoeuvres of flocks of starlings on their way to roost, or the habit of monarch butterflies of overwintering in densely packed groups on one or two individual trees.

Wynne-Edwards argues that behaviour serves as one link in a chain of responses, not just behavioural but also physiological, which match the breeding population to the food supply. Thus communal displays will enable individuals to get some measure of their density. If this is high, breeding is reduced in the subsequent season; if it is low breeding increases. Now of course he is not suggesting that there is any conscious self-sacrifice or adjustment on the part of the animals. Rather selection has favoured groups in which such adjustment mechanisms operate automatically. We know for example that titmice regulate their reproductive rate according to the available food supply (see Lack[143]). In years when the caterpillar crop is low they lay fewer eggs than in good years. This response is 'anticipatory' because the birds lay their eggs before the caterpillars hatch. Such a remarkable adjustment must result from the birds' physiology being matched to climatic or other environmental factors which also affect the reproduction of the moths. A similar kind of response could evolve which matched reproductive physiology to the density of a social display (we have just described how the 'Fraser Darling effect' links social behaviour and the timing of reproduction).

Thus in principle there is nothing improbable about Wynne-Edwards suggestion on the rôle of social behaviour. The controversy that surrounds his theory is largely concerned (i) with the concept of group selection versus individual advantage and (ii) with his relative neglect of the other factors which might lead to the evolution of displays and social life, some of which we discussed earlier in this chapter.

There are certainly great difficulties with the concept of group selection. Suppose, as Wynne-Edwards suggests, a colony of birds lives within its food supply by laying fewer eggs than the maximum. In fact many sea birds do lay only a single egg and do not replace it if they lose it. Clutch size must have an hereditary basis and if there is enough food in the short-term, how could selection prevent the spread of genes which caused birds to lay 2 eggs? Even if we assume the 1-egg-only convention could evolve—and it is hard to see how this happens unless it benefited not only the colony but the individuals—it would seem to be an inherently unstable situation because new mutations would constantly threaten it in the face of short-term individual gains.

We cannot go further into the controversy here; something of its intriguing style and flavour can be gained by reading the appendix to Lack's book,[143] and articles by Wiens,[281E] Crook,[55A] Smith[245A] and Wynne-Edwards[287A] himself.

The idea that animal populations are regulated to live well within their resources is an attractive one, but it must be admitted that such evidence as can be mustered in its favour is largely circumstantial and inconclusive. A few seasons' work on the behaviour and reproductive biology of a population is not sufficient; it may well take 25 years to get a reasonably reliable answer. However there can be no doubt of the service Wynne-Edwards has rendered by drawing attention to possible connections between social behaviour and population control. This forces us to think of sociality as an adaptive feature which transcends individual responses—although dependent upon them—thus enabling a higher order of adaptiveness to emerge. In other words adaptation can involve not just a species becoming social, but also the nature of its social organization.

SOCIAL INSECTS

The form of social life exhibited by some of the insects is in many respects so distinct from that of other animals that we can justifiably treat them separately. Nevertheless, for all their distinctiveness, in two key aspects of social life—division of labour among the members of a society, and communication between them—the social insects exemplify principles which can be applied to all societies.

Amongst the insects we find some of the most completely solitary animals. This is because they may lack even the most basic of social contacts, that between parent and offspring. The short life span of many insects, coupled with seasonal reproduction, often means that parents have died long before their offspring emerge from larval life. The digger wasp whose life history was given in outline in Chapter 2, p. 17, has only one brief contact with another member of her species when she mates; apart from this she is totally solitary.

It is obvious that one essential for the development of social life is to lengthen the life span so that there is contact between old and young individuals. The aggregations formed by insects such as cockroaches and earwigs contain such overlapping generations. The life span of a cockroach may be a year or more, three quarters of which is development when it passes through a series of nymphal stages, becoming gradually more like the adult insect. During this period cockroaches of all ages live together in a loose aggregation near sources of shelter and food. Some cockroaches incubate their eggs inside the female's body and bear live young which remain in contact with the mother for some hours after birth. This contact may be important for the survival of the young because, not only are they extremely vulnerable to cannibalism when first born, but they may pick up nutrients from the mother's body surface.

We know that this nutritional factor is of major importance in the termites (which are related to the cockroaches) because, feeding largely on wood, they rely on symbiotic protozoa living in their gut to digest cellulose. These protozoa are acquired by the young termites when they feed on fresh faecal matter from the adults, and they can be transmitted only in this way. Perhaps this essential overlap between the generations was one basis for the evolution of the termites' elaborate social life.

True societies with organized structure are found in two orders of the insects, the Isoptera or termites as just mentioned and the Hymenoptera, the ants, bees and wasps. There is a huge literature on social insects which have attracted man's attention for centuries. Here we can only give a very general account of some of their characteristics which are most relevant to a discussion of social organization. For more details, and for an introduction to the whole literature, consult Wilson[283B] for a complete modern survey, books by von Frisch,[77] Lindauer,[153A] Butler[42A] and Sudd,[251C] and a useful series of review papers edited by Ellis.[69B]

The unique feature of an insect society is that although it commonly contains thousands of individuals, they are all closely related to each other and constitute, with certain minor exceptions, a single family. Reproduction is generally confined to one female (the queen) and the other members of the society are her offspring, which remain sterile. Living in a common nest which they construct, they assist the queen to produce more workers like themselves and eventually more reproductive castes.

This term, caste, is well suited to describe the division of labour within insect societies. It implies a rigid, limited rôle in society largely determined by one's upbringing. Certainly the most important factor determining caste is diet. In bees, wasps and termites all eggs laid by the queen are potentially equal, but most larvae are fed a restricted diet and develop into workers. There is evidence that when queen ants are laying rapidly their eggs are 'worker biassed' and develop accordingly no matter how the larvae are fed. But for most of the time their eggs are also equipotential and only the richly fed individuals develop into the reproductive castes. Termites have both males and females represented in all castes (eg. there are 'king' termites, which remain with the queen and mate with her several times during her egg-laying period), but in the Hymenoptera all the workers are female. The males (or drones as they are called in bees) constitute a separate caste. They are produced at the same time as new queens, and the sex determination mechanism in Hymenoptera is such that males always develop from unfertilized eggs and are thus haploid whilst the females and workers are diploid.

In most social insects, with the notable exception of the honey-bee which reproduces colonies by swarming, new colonies are founded by a single queen (or pair in termites). She begins the construction of the nest and rears the first batch of workers herself. These then take over the tasks of extending the nest and bringing food, and the queen usually stays in the nest laying eggs from this point on. In termites the gradual growth of the young through the various nymphal stages, as in the cockroach, mean that it can play its part as a working member of the colony right from the start. In the Hymenoptera the helpless larvae require protection and feeding by adult workers until following metamorphosis they themselves emerge as fully-grown adults.

The tasks performed by the worker castes vary greatly in detail, but in most colonies they cover the main categories of foraging, rearing the young, nest constructions, attending the queen as she moves about laying eggs and guarding the colony. In termites and ants this last task is sometimes the sole responsibility of a special soldier caste which has enlarged jaws or other weapons. Less specialized workers carry out a range of the other duties with varying degrees of attachment to any one task. In bumble-bees very small workers are sometimes produced because of poor nutrition as larvae, and these tend to stay in the nest all their lives, never going out to forage. Food supplies are much more regulated in honey-bee colonies, workers are all of one type and perform all the tasks listed above. Division of labour in the hive is based upon the age of the workers, but it operates quite flexibly.

A honey-bee worker lives for about 6 weeks as an adult and her activities are to some extent synchronized with her physiology. Thus she spends the first 3 days cleaning out cells, and then begins feeding the older larvae on a mixture of pollen and honey which she picks up from the storage cells in the hive. During this period the pharyngeal or 'nurse' glands in her head have been developing. They secrete the so-called 'royal jelly', and from about the 6th to 14th day of her life the worker feeds this secretion to the younger larvae and any queen larvae in the hive. (Royal jelly is fed to all larvae for a brief period early in their development, but those larvae intended to become queens develop in a larger cell and are fed royal jelly throughout.) The worker's wax-secreting glands on the abdomen become active from the 10th day and at the same time the pharyngeal glands begin to regress. She gradually changes her behaviour from feeding larvae to cell construction. From about the 18th day she may leave the hive occasionally for a few brief orientation flights (see p. 191). At this age she may be found guarding the hive entrance and inspecting incoming bees. From 21 days of age onwards the worker is primarily a forager, bringing back nectar pollen and water, and usually remains so for the rest of her life—2 to 3 weeks.

This is the general sequence, but it can be modified to suit the needs of the colony which are certainly not fixed but vary with the flower crop, the temperature, the age of the colony and many other factors. It can be modified, because of the remarkable communication system which exists between the members of a honey-bee society.

Fig. 10.2 shows a record kept by Lindauer[153A] of the activities of one individual worker throughout her life. The sequence just described can be discerned, but the most conspicuous activities are two not previously mentioned, 'resting' and 'patrolling'. Clearly one way in which the individual is made aware of the needs of the colony is by personal inspection. The worker moves about the hive exploring empty cells, the food storage areas, the edges of comb where new construction is proceeding and the brood area where the larvae are developing. In this way she is stimulated to initiate activity of her own. If there are a large number of underfed larvae, the pharyngeal glands of an older forager regenerate and she moves back to feeding. If food is short, young workers prematurely become foragers.

The most obvious way in which information on the state of the colony is transmitted involves contact between the workers. A resting bee is easily aroused by the activities of other workers nearby. There is incessant contact between bees on the combs and incoming foragers are approached and solicited for food. It is not just from foragers that food is taken, any pair of workers that meet may proffer or solicit food. One bee extends its tongue and rubs its antennae along the antennae of the facing worker, who then regurgitates a drop of nectar between her mandibles. In a second or two this is lapped up and the two bees move on. Within a few moments the receiver may become the donor as she, in turn, is solicited or proffers food to another.

Nixon and Ribbands[199B] fed a small quantity of sugar solution containing radioactive phosphorus to 6 foraging bees from a colony of 24,600 bees. Within 5 hours radioactivity could be detected in 62% of foragers and 18% of bees tending brood. After 29 hours 76% of foragers and 43% of those in the brood chamber were radioactive. Thus within a day the food collected by 6 bees had been shared between many thousands.

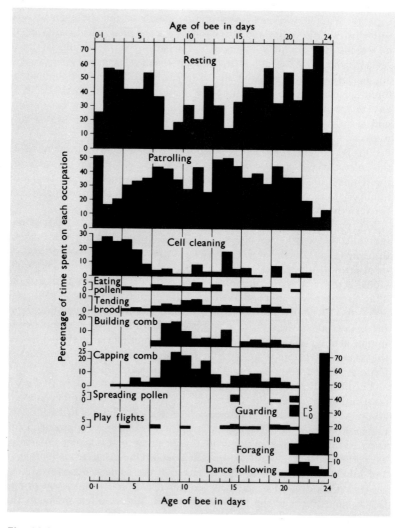

Fig. 10.2 Lindauer's complete record of the tasks performed by one individual worker honey bee throughout her life. The records are classified according to the type of task. One can recognize the age-determined succession of cell cleaning, brood care, building, guarding and foraging. Note, however, the large amount of time spent in patrolling the interior of the hive and in seeming inactivity. (From Lindauer,[153A] 1961, 1971. *Communication among Social Bees.* By permission Harvard University Press, Cambridge, Mass. © President and Fellows of Harvard College.)

Such incessant food transmission is basic to communication in all insect societies (Fig. 10.3 shows the analogous behaviour of ant workers). Not only is every individual kept directly informed of the state of food supplies in the colony, but pheromones (see p. 136) are also circulated. The development of worker termites is controlled by pheromones produced by the king and queen. Similarly the queen honey-bee secretes a pheromone

Fig. 10.3 Food sharing between workers of the ant, *Formica fusca*. The worker on the left is receiving food from that on the right which has regurgitated a drop of liquid from its crop and offers it between the outstretched mandibles. (From Wallis, D.I. 1962. *Animal Behaviour*, **10**, 105–11.)

('queen substance') which both suppresses the ovaries of the workers and prevents them from rearing new queens. The queen is always surrounded by attendant workers who lick her body, subsequently proferring food to other workers. The level of the pheromone must be kept up and once the source is cut off its concentration rapidly drops below the critical level. The effectiveness of the incessant food sharing in circulating queen substance is shown by the fact that some workers in the brood area of the hive exhibit changed behaviour within an hour or two of a colony losing its queen. They begin construction of an 'emergency' queen cell in which one of the youngest larvae, destined in the normal course of events to become a worker, is fed royal jelly throughout its larval life and becomes a queen to replace her mother.

In nearly all the social insects there is further communication between the foragers about sources of food. In ants and the group of 'stingless' bees (Meliponinae) communication is by purely chemical means.[283A] When leaving a food source to return to the colony, ants lay scent trails and the Meliponines dip down from their foraging flights at intervals of a few yards to mark leaves or the ground with scent from special mandibular glands. Subsequently other foragers are attracted by such trails and can be shown to find the food more easily.

We have already discussed the evolution of the 'dance language' of honey-bees. This is certainly the most remarkable example of communication between social insects, because of the detailed information which it conveys about the distance and direction of a food source. One vital factor in the operation of the dance is the close attention paid to incoming foragers by other workers. As she moves up on to the comb the forager is contacted many times and she can gauge the needs of the colony by the eagerness with which her preferred nectar load is accepted. The persistence of a forager's dance on the comb is related to the richness of the food source which she has discovered. Accordingly other foragers who may be resting in the hive are more likely to be attracted towards rich sources. On the other hand if her food source was not a rich one, the incoming forager may not dance. She is quite likely to stay in the hive and subsequently be attracted by the dances of more successful foragers.

Although other social insects lack the detailed communication system of the honey-bee, their foraging is not at random. As we mentioned, incoming foragers may lay scent trails and other workers become aroused to go out and forage by the highly excited behaviour shown by a forager who has discovered a rich food source. Thus social insects tend to forage as a team which directs its resources where they will be most profitable.

Throughout this brief survey of social insect behaviour we have used descriptive terms which may give the impression that their life is organized in an intelligent way. We must not confuse adaptiveness with intelligence here. The beautiful adaptiveness of the social insect colony and the control it gives them over their environment are based upon a relatively simple series of responses to other workers in the colony and to the nest itself. Within the limits of their remarkable capacities the social organization of the insects is flexible, as when honey-bee workers change their normal sequence of tasks in response to a sudden requirement in the colony. However the type of response the social insects can make is itself determined within inherited limits which are characteristic of the species. All honey-bees respond the same way, so do all ants of a particular species and so on.

This consistency of social organization within a species is highly character-istic of the insects and we shall not find it so well marked in other animals.

SOCIAL ORGANIZATION IN VERTEBRATES

Very few vertebrates are as solitary as the digger wasp; they live far longer and often have well developed parental care, both factors which ensure that there is overlap of the generations. Nevertheless they exhibit a very wide diversity of social organization and furthermore, in striking contrast to the social insects, such organization is not rigidly species speci-fic. Thus the answer to the question, what is the social organization of the house mouse? is not fixed. It depends upon the density of the population, its age and sex structure and a number of other factors.

Recently there have been a number of important surveys of social be-haviour in diverse animal groups which try to relate social organization to the whole life and ecology of the species (see in particular Crook[55A, 55C] and Itô[130]). Particularly amongst the higher vertebrates it is possible to detect a pattern which, as mentioned previously, reveals social behaviour within a group as a higher order adaptation to the environment—one which goes beyond the adaptive responses of the individual.

Some features of social organization are, at least in part, imposed by the environment itself, notably the size of the groups in which animals live. Often the group size of birds, ungulates and primates living in open grass-land or plains resemble one another more than each does its own rela-tives living in forest habitats. In open habitats communication is easy, groups wander quite widely and tend to be large. In dense forest a large group would have formidable problems in maintaining its cohesion; accordingly groups tend to be small and many forest animals are solitary. The large groups of hippopotamus in open lakes and rivers and of red deer in the Scottish Highlands can be compared with the solitary pigmy hippo-potamus and the moose of dense forests. Even within a species we can sometimes see the same trend. Prides of lions rarely amount to more than a family party of 6 to 8 when they are living in thick bush, but in open savannah groups of 30 are not uncommon. Of course as with any biological law there are exceptions, but such overall trends show that it is profitable to look for the adaptive features in different types of social organization.

As indicated, there has been a remarkable increase in studies on verte-brate social organization. Many such studies have been made on wild populations in the field. Partly because of this and partly because of the nature of social behaviour, the concepts and terms used in describing it are sometimes less precise than those that can be applied in experimental studies and this will become evident from what follows in this chapter. This is inevitable at this stage of analysis where rather little is known of social organization in most vertebrates and we are still at the pre-experimental 'comparative anatomy' stage—just describing what happens without trying to change it. As Crook[55C] points out we shall need to go beyond the conventions of ethology, which evolved largely to cope with the behaviour of individuals. The new breed of socio-ethologists will have also to use concepts borrowed from sociology, for example.

In our general discussion of vertebrate social systems we shall try to bring together information from a wide range of studies. More detailed discussion and fuller references to the literature can be found in Etkin[70A] (a general survey), Crook[55A] and Wood-Gush[285] (for birds) and Chapter 4 in Ewer's book[71] (for mammals).

Territory and dominance hierarchy

The rôle played by individuals within most vertebrate societies depends on their age and sex, but it is far less rigidly determined than in the social insects. McBride,[163] in an important discussion of social organization based on studies of the fowl, uses the term 'caste' to represent the different stages specialized in 'function, behaviour and way of life' through which the bird passes. But the very fact that a bird's caste changes as it grows up, and between seasons when it is adult, shows that caste cannot retain its complete insect meaning. McBride's aim is to direct attention to the relatively proscribed nature of a bird's rôle at any one stage. In the majority of vertebrates, as in the fowl, the individual's rôle and the interactions it makes with other members of the social group involve assertions of rank or dominance. Territory and hierarchy represent in some ways two extremes of such organization.

We have discussed various aspects of territory before, especially in Chapter 6, and at first sight territoriality may seem to be the antithesis of social behaviour. Territory takes many forms, and that illustrated for willow warblers in Fig. 5.1 represents a fairly typical case. Each male defends a substantial area which will include an adequacy of food for himself and eventually a mate and young, so that he will rarely leave his territory. It impinges closely on the territories of others and, where the habitat permits, nearly all the ground is occupied. Territories of this type are also found in some mammals, notably carnivores, where territory is defined by scent posts marked with urine or special glandular secretions.

Whilst the territory holders are largely solitary, the whole group of territories certainly constitutes a social organization if not a society and the fact that territories are grouped will, for example, influence the behaviour of females seeking mates. In sea birds, such as gannets, gulls and terns, this grouping is much more obvious. The territory is no longer for feeding but simply a small area around the nest. The territories are tightly packed and the birds form a coherent colony.

The common feature of all territories is that they confer complete dominance on the owner within the boundaries. There he is secure and the boundary of his territory marks the point at which his dominance begins to give way to that of his neighbour. The hierarchical organization of dominance does not relate to a fixed area, but to rank order between a group of individuals living in a common area. The group will move around together and may indeed defend a communal territory, but within the group some animals are dominant over others. This means that they can displace more subordinate individuals for food or a mate, for example.

Schjelderup-Ebbe developed the concept of hierarchy from his work on flocks of birds (see his summary[231]). He observed that a definite 'peck-order' developed amongst small groups of chickens confined together in a pen. As the birds disputed amongst one another one gradually emerged as the dominant who could displace all the others. Below her was a second-ranking bird who could dominate all except the top bird and so on down the group until at the bottom was a bird displaced by every other in the flock. Fig. 10.4 shows one example of a peck-order of this very precise type —a linear hierarchy as it is sometimes called. The hierarchy develops as the birds dispute and involves a good deal of fighting in the early stages, but once established it is as much an hierarchy of submission as one of dominance. Subordinates usually defer without question at the approach of a more dominant bird. Note that individual recognition is a prerequisite of stable hierarchies and Schjelderup-Ebbe found that they failed to form in large flocks where disputes continued as recognition failed.

Hierarchical structure of same type is very widespread amongst the vertebrates and in some invertebrates too, but it rarely assumes such a perfect regularity as that shown in Fig. 10.4. Divergences from the 'pure' type can take several forms. The hierarchy may not be linear, and triangular relationships may develop where A displaces B who displaces C, but C can displace A. In primate groups it is not uncommon for high-ranking animals to co-operate to assert their dominance over others. This may make it difficult to separate them in the hierarchy. Again, sometimes it may not be possible to distinguish any ranking below the top animal, i.e. one animal dominates all the rest who can only be lumped together as 'subordinate'. This is often the result if male mice are kept in a fairly crowded group. It also occurs if male sticklebacks are placed in a tank with inadequate space. The first one to develop into reproductive condition usually takes the whole space as his territory and all the others are driven into submission.

Territory and hierarchy are obviously not completely distinct types of social organization—as we shall see some systems which might be called territorial-hierarchical do occur—but they are useful descriptive terms around which to organize our discussion.

Perhaps the simplest social structure is that of solitary animals in which both sexes hold a separate territory, whose boundaries break down only to allow reproduction. Hamsters are strongly territorial and the female will only allow a male to approach her for an hour or two at the very peak of

	Y	B	V	R	G	YY	BB	VV	RR	GG	YB	BR
Y												
B	22											
V	8	29										
R	18	11	6									
G	11	21	11	12								
YY	30	7	6	21	8							
BB	10	12	3	8	15	30						
VV	12	17	27	6	3	19	8					
RR	17	26	12	11	10	17	3	13				
GG	6	16	7	26	8	6	12	26	6			
YB	11	7	2	17	12	13	11	18	8	21		
BR	21	6	16	3	15	8	12	20	12	6	27	

Fig. 10.4 A perfect linear hierarchy established within a group of twelve hens. Each bird is marked by colour rings on its legs whose initials identify it. The number of times each bird pecked another flock member is given in the vertical columns (e.g. Y pecked B 22 times and V 8 times) whilst the number of pecks received from another is given in the horizontal rows, (e.g. VV received 19 pecks from YY and 8 pecks from BB). Note that no bird was *ever* seen to peck an individual above it in rank : hierarchies as perfect as this are probably rare in nature (From Guhl, A. M. 'The Social Order of Chickens.' *Scientific American.* © 1956 by Scientific American. All rights reserved).

oestrus. The European robin also has territory in both sexes and female robins, like males, sing throughout the winter. During spring female song subsides and pairing takes place, usually on the male's territory which both birds then defend.

Leyhausen[152A] has described an interesting variation of territory amongst solitary cats. The domestic cat has a fairly well defined territory with certain areas which are marked with urine, linked by a network of favourite paths. However the territories are not held exclusively; neighbouring cats avoid one meeting another by using common areas at different times. They learn one another's habits and this apportioning of access to an area by time breaks down only when males compete for an oestrus female. The territory of a cat would normally provide an adequate food supply in the wild. It is of great interest that Eaton[68] has recently found a similar time-sharing system in the African cheetah. Small groups of cheetahs, probably derived from a family and sometimes containing more than one male, move about together. The males scent mark frequently and examine scent posts for traces of other cheetahs. If the scent is fresh the group alters its direction of march, but it continues if the scent is more than about 24 hours old. In this way meetings between groups are reduced to a minimum and available space is shared efficiently.

Amongst some birds and mammals not all individuals hold territories of the same type. There is a combination of hierarchy and territory such that dominant animals hold larger territories or hold them in the most preferred areas. Familiar examples are provided by the 'lek' system of birds such as the black grouse and the ruff. The lek is an area of ground divided into a cluster of territories held purely for mating. Males compete for desired positions near the centre of the lek and here territories are small. Less successful males may hold large territories at the periphery of the lek, but females are attracted by the displays and high activity of males in the central areas and it is these males who have most success in mating.

Fig. 10.5 shows a map of territories on a black grouse lek with 5 small territories at the centre and a number of large territories at the periphery whose boundaries are sometimes indistinct. Kruijt and Hogan[142B] found that central males arrived first at the lek soon after dawn each morning and stayed for several hours. These males probably kept the same territories from season to season, but eventually were replaced by younger birds from the periphery. Each year some young birds became established at the periphery but a male was usually several years old before he could hold a central territory. Visiting females tend to alight in the territories of central males, who display vigorously to the sight or sound of grouse flying over the lek. Thus the younger peripheral males are largely excluded from mating, although if they are persistent and return to the lek each season, their turn will come. Once she has mated, the female grouse leaves the lek area to nest and rear the young alone. The male's rôle is simply to attract as many females as possible and in black grouse, ruff and other lek birds, just as with other polygamous species discussed in chapter 7 (p. 172), males tend to have elaborate plumage and be larger than females.

The lek displays of birds have been known for centuries, but recent work has shown that some mammals have a very similar type of social organization. Fig. 10.6 shows a map of territories held by males of the kob antelope of East Africa. In each local population of kob, which may number around 15,000, a favoured area of land is occupied by adult, territorial males. Younger males and females move about in small bachelor herds as do females with their young. As males become sexually mature they tend to leave the herds and become territorial, especially during the dry season

Fig. 10.5 Territories on a black grouse lek on pasture land in the Netherlands observed during 1964. The two vertical black lines represent the position of fences which are sometimes adopted by the birds also as territorial boundaries. Crosses mark uncertain boundary lines. There is a central area occupied by cocks A, B, C, D and G on small territories. The peripheral birds can occupy much larger areas. (From Kruijt, and Hogan,[142B] 1967. *Ardea*, 55, 203.)

when grass is short and does not restrict movement or visibility. As can be seen in Fig. 10.6, territories vary greatly in size. Some—which Leuthold[150A] calls 'single territories'—are 1–200 metres in diameter, others are much smaller. The smallest territories are clustered tightly together to form a 'territorial ground'. Here each male is defending an area less than 30 metres across and there are frequent disputes and various advertising displays. The ground is trampled flat and bare.

Males may occupy single territories for weeks and, since they can get a good proportion of their food on the spot, spend most of their time there. There is not usually much competition for these large peripheral territories but competition becomes much stronger as one approaches a territorial

0   100   200 metres (approximate scale)

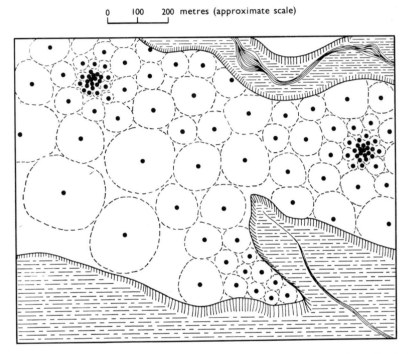

Fig. 10.6 Territorial map of males of the Uganda kob on an area of grassland raised above two swampy river bed areas, indicated by shading. The boundaries are only very approximate, but each black dot represents the centre of a territory, close to where a male usually stands. There are two closely packed 'territorial grounds' and a third may be forming on the 'peninsula' at the lower part of the map. (After Leuthold, [150A] 1966. *Behaviour*, **27**, 215.)

ground. There is quite a rapid turnover of males in these small territories in part, at least, because lack of food and rest forces males to leave. Males leaving the territorial areas for any length of time rejoin all-male herds in the vicinity·

Female herds move through the territorial areas, and those in oestrus detach themselves and approach males. Although they are courted by males on single territories, almost all copulations occur with males in the territorial grounds. The situation is therefore strikingly analogous to that on the black grouse lek. If the analogy is to be exact, then single territories should be held predominantly by young males who would gradually win their way into the central group, but this does not necessarily seem to be the case. Leuthold found that the age structure of groups of males holding single territories was the same as that for males from the territorial ground. He suggests that the territorial group phenomenon is related in some way to a high population density of kob. In some other areas where kob are less abundant, single territories seem to be the rule and oestrus females mate with males holding them.

Certainly there are, as we suggested earlier, numerous examples of changes in social organization with density. Often crowding leads to a more hierarchical type of structure. Experimental studies on animals as diverse as sunfish (Erickson[69E]) and mice (Davis[60D]) have shown that males form typical hierarchies under high density but lapse into territoriality when space and cover permit. Owen-Smith[202B] suggests that it is high population density thanks to protection which—for once—has worked, that accounts for the rather anomalous territorial structure of white rhinoceros in a South African game reserve. Adult bulls are strictly territorial, but many territories now include a 'subsidiary' bull, who always defers to the territory holder and whose presence is largely ignored. This may be one stage in the breakdown of the territorial system to form an hierarchy. Since territories are normally about 2 km^2 and there are now about 5 rhinos (of all ages and both sexes) per km^2 there is obviously pressure on the normal social structure.

Territoriality and density are not always related in this way—everything depends on the general ecology of the species concerned. Estes[70] finds that wildebeeste tend to be more territorial in areas of rich food supply. Here densities are high, but the paramount factor is adequate food which allows the animals to be sedentary. Male wildebeeste always defend territories for breeding but in poorer areas these are only transitory. For much of the time the animals have to be on the move, following grass where it is to be found. They form large herds, usually consisting of a mixed group with associated all-male herds, from which latter males detach themselves to take up temporary territories as conditions permit.

For many vertebrates the environmental changes which they meet with changed social organization are the regular, seasonal ones. A wide range of birds which are territorial in the breeding season form flocks in the winter, sometimes combining with other species. The selective advantages of moving and feeding as a group were discussed earlier. During the more rigorous winter conditions these outweigh the advantages of a territory for each individual, whose selective value is largely associated with reproduction and rearing the young. With most passerine birds there is little social structure within the winter flocks. They move together, respond to one another's flight and alarm calls and roost together, but this is all. Chaffinches do have a tendency to form one-sex flocks and we know from the experimental work of Marler,[178C] that males tend to dominate females in winter and displace them at feeding sites. There is a reversal of this dominance in spring and females tend to displace males when the flocks break up and males disperse to set up territories.

Some geese and ducks retain family groups at least during migration and perhaps in the winter flocks. The fowl, studied in a truly feral state on an island off Queensland, Australia, alternates between a territorial system in the breeding season and a more hierarchical flock structure during the winter. McBride et al.[163] found that only the dominant or alpha male of each flock sets up a territory in spring. He mates with several of the hens from his flock during the season. During winter after the young birds of the year have returned to the flocks, the alpha male and his harem move about over a home range with numbers of subordinate males staying at the periphery of the group, often moving between the home ranges of different alpha males. The alpha male leads his flock in every sense. He it is who initiates all movements of the group, particularly across open ground, and his posture is normally more alert than those of his females. He is the first to give alarm calls and may even approach predators—feral cats— whilst the others take cover.

Seasonal changes in behaviour are just as marked in some mammals. The classic study *A Herd of Red Deer* by Fraser Darling[60B] remains one of the most complete of any ungulate species. The sexes normally remain separate when adult and the female herds, which include young animals of both sexes, move within a preferred home range which in summer extends on to the higher feeding grounds. There will usually be one dominant old female who acts as leader. The herds of stags are smaller and may be loosely associated with the female herds. In autumn the male herds break up and the stags, their antlers fully grown, become highly aggressive, roaring and disputing between themselves. Oestrus females are attracted to displaying stags who gather a harem, and they defend their group of females rather than any particular area. The rutting season is relatively brief and soon the females return to their herd for the winter move on to lower ground. The stags shed their antlers and once more join together.

The striking changes in social behaviour which accompany the onset of the breeding season in the vertebrates we have hitherto been describing, are less marked in those species where the group co-operate to hunt and to defend themselves. Wolves, for example, hunt in packs and maintain a very stable social structure based on an extended family unit. There is usually one dominant male leader, but several other adult males may be included. Primates also tend to retain a uniform social structure through-

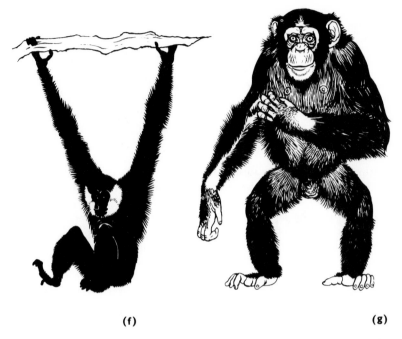

(f)   **(g)**

Fig. 10.7 Some diverse living representatives of the primates. Although they do not in any way represent a phylogenetic series, the higher primates and man did have ancestors which were rather similar to their more primitive relatives illustrated here. (a) A tree shrew (*Tupaia*)—probably best regarded as a specialized insectivore rather than a primate, but certainly resembling the stock from which the early primates arose ; (b) the tarsius (*Tarsius*) of the Philippines ; (c) the ring-tailed lemur (*Lemur*) of Madagascar ; (d) a South American capuchin monkey (*Cebus*) ; (e) the pig-tailed macaque of S.E. Asia (*Macaca*) and two of the apes—man's closest relatives—(f) the gibbon (*Hylobates*) and (g) the chimpanzee (*Pan*). (After Le Gros Clark,[53] 1965. *History of the Primates*, 9th edn. Brit. Mus. (Nat. Hist.), London, not drawn to scale.)

out the year and in conclusion we may turn to consider their social behaviour in more detail.

Primate social organization

Studies on the primates have accelerated more rapidly than those on the social behaviour of any other type of animal. Workers from a number of disciplines, zoologists, psychologists, anthropologists and sociologists have converged on this group, both for their intrinsic interest and in the hope that they will provide information which is relevant to speculations on the origins of human societies. The great emphasis in the recent primate work has been on studies of natural communities in the field. Certain primates, notably the rhesus monkey, have been familiar laboratory animals for some time, but it is generally accepted that studies on captive communities are inadequate by themselves.

Here we can attempt only a brief survey of the primate work. There are now two journals devoted mainly to behavioural studies, *Primates* and *Folia Primatologia*, but for an introduction, several recent symposium volumes offer a good variety of review papers: these are those edited by Altmann,[4A] De Vore,[66A] Jay,[131B] Morris,[196A] Napier and Napier[198A] and Southwick.[248A]

Primates live in a wide variety of habitats. We perhaps tend to think of them as tropical animals, but they were more widely distributed in the recent past and two of the macaques, the Barbary 'ape' of the Atlas mountains and the Japanese monkey live in areas where snow and frost are regular every winter. The majority of primates are arboreal, some of them like the spider monkey of South America and the colobus monkeys of Africa, exclusively so. However a number have returned to ground living, such as the baboons and the patas monkey of Africa, whilst the chimpanzee and the gorilla also spend a lot of time on the ground. In general the more arboreal primates are fruit and leaf eaters, the ground dwellers tend to be more omnivorous. Although the gorilla is exclusively vegetarian, chimpanzees and baboons will eat meat if they get the opportunity; the former have been seen to hunt and kill monkeys for food.

The living primates exhibit a tremendous range of morphological types, from the primitive lemurs which retain a long muzzle, a moist nose and claws on one of their hind toes—little modified from the ancestral primate types—up through the monkeys to the great apes and man (see Le Gros Clark[53] and Fig. 10.7). Throughout this series we can observe certain trends; the enlargement of the brain, the development of the grasping hand and in contrast to many other mammals, the great reliance on colour vision as a dominant sense for exploration and communication.

It seems certain that from very early in their history the primates were social animals moving around in groups whose organization was stable. Jolly[132A] makes this point from her study of lemurs; here in these primitive primates we already find small mixed troops (12-20 individuals) which include several adult males—a very typical primate grouping. There is a dominance hierarchy within the troop (and in some lemurs females rank more highly than males) but it remains as a permanent, cohesive unit. Lemurs have group territories within their mixed woodland habitat whose boundaries are often remarkably stable. They are marked by scent in some species—and are defended by calling, which is usually sufficient to cause the neighbouring troop to retreat without further threat or fighting.

Within the troop there are frequent minor disputes but serious fights are rare outside the breeding season. This is very brief in most lemurs— at most two weeks—and it is at this time that subordinate males seriously challenge the older dominant ones. Apart from this short period of strife, much of lemur social life is characterized by non-aggressive interactions between individuals—indeed it is pedantic to avoid use of the term 'friendly'. There is always close contact between a mother and her infant who clings continuously to her at first and is carried around everywhere. As it grows older other adults approach and play with the infant, as they also play with each other. Lemurs have thick, dense fur and groom frequently. Mothers groom their infants and adults frequently groom each other—this being one of the commonest types of friendly contacts between individuals.

In the behaviour of lemurs we can detect most of the elements which characterize all primate societies, although there are many variations on the theme. These variations involve group size, territoriality and intra-group relationships to mention some of the more obvious ones.

Some of the smallest groups are found in two of the apes—the gibbon and the orangutan. The orang of Borneo and Sumatra appears to be an exceptionally solitary animal; it is rare to see more than 2 or 3 together moving through the high canopy of tropical rain forest and lone males are common. Next to nothing is known about its social interactions but they are certainly sparse—it remains something of an enigma for an animal of such high intelligence. Gibbons live as small family parties, a monogamous pair and their young offspring. They are territorial and hoot loudly when they detect other gibbons in the vicinity: usually a contest of calls is sufficient to maintain the boundaries.

Gibbon and orang both live in thick forest where, as we mentioned at the outset, group size in any vertebrate tends to be small. However, other dense forest primates such as the howler and several other South American monkeys live in moderate sized groups, 12–30 or so. Often too they tend to avoid other groups and have their own preferred home range but unlike the gibbons, they cannot really be said to defend a territory. In a large home range in forest it is impossible for a group of monkeys to survey and patrol boundaries in the way—say—a bird can on its territory. As Carpenter[44A] says of howler monkeys, '(they) do not defend boundaries or whole territories; they defend the place where they are, and since they are most frequently in the familiar parts of their total ranges, these areas are most frequently defended—typically by interchanges of roaring at approaching or approached animals.' Other forest monkeys have analogous calls which serve to space out their groups. Marler[179A] describes the calling behaviour of the African colobus monkey. He concludes that there is a true territory here in the sense of an actively defended area, but the colobus also roam beyond this at times. Marler concludes that above about 1 square mile of home range, territoriality in the usual sense disappears and inter-troop contacts must become fairly rare in any case.

Inter-group avoidance seems to be less marked in the more terrestrial primates, although it is difficult to make generalizations. Certainly there can be a considerable amount of overlap between the home ranges of Indian langurs, baboons and chimpanzees for example. Hall and De Vore[93C] report several instances of two groups of baboons intermingling during food foraging with no sign of hostility. Some of the largest groups are to be found in the terrestrial primates—baboon troops range up to 90 strong and the baboon-like gelada of the high Ethiopian plateau sometimes has herds of 400, although there are many social sub-units within a party of this size.

One cannot fail to be impressed and fascinated by the high level of inter-communication that goes on between the members of a primate group. Each individual is constantly responsive to the movements, gestures and calls of others. Naturally, we must be cautious about assumption that the primate societies are more organized than those of—say—ungulates or carnivores. Because we ourselves are primates we find it much easier to identify elements in their communication system. In particular the mobility of their faces and the way they watch one another's faces for information on mood and intentions. Nevertheless there are, on the most objective criteria, good grounds for the assumption. Two factors that contribute to this situation are the high learning ability of primates and their extended period of infancy, to which is coupled considerable longevity. The larger primates commonly live for 30 or 40 years and this means that a young primate grows up to take its place in a group where—literally—everybody knows everybody else from long experience in their company. (There are several good reviews of both visual and vocal communication in Altmann[4A] and Morris[196A]).

Primate societies usually involve a stable hierarchy of dominance. Normally the hierarchy is reasonably linear and in rhesus macaques and baboons, for example, all adult males rank higher than females. Having high rank in the group determines access to food, preferred sleeping places and to females. When they come into oestrus females tend to form a brief 'consort relationship' with the most dominant males, and move close to them. Rank is often determined initially by threat and fighting, but once established it is maintained as much by the deference of subordinate animals as by any display of threat by the dominants. In fact overt aggression is not common under normal circumstances and there are many friendly contacts between animals as when they invite grooming or offer to groom another. Mutual grooming, which we first mentioned in the lemurs, is very important as a placatory gesture in primates (see Plate 4). Often a dominant animal will 'allow' itself to be groomed by a subordinate follow-

Plate 4 Friendly grooming in the Barbary macaque. The juvenile male on the right is about 3 years old ; he is grooming a 'subadult' male aged about $4\frac{1}{2}$, who ranks above him in the social hierarchy. A 10 week old baby is being held by the subadult male. (Photo by John Deag.)

ing a brief threat to which the subordinate has deferred. Sexual presentation as an appeasement gesture was discussed briefly in Chapter 5, p. 111; it is very common in baboons and chimpanzees and is made by males or females towards a dominant animal who threatens, or even if the subordinate wants to pass close to the dominant one, see Fig. 10.8.

The rôle of aggression in primate societies has attracted a great deal of attention. As may be imagined, the primates have been used as evidence by both sides in the dispute about the nature of human aggression which was

Fig. 10.8 Sexual presentation in the chacma baboon, (a) 'genuine' presentation by an oestrus female, she characteristically looks back over her shoulder towards the male, (b) appeasement presentation towards a threatening dominant animal, the general stance is the same but the facial expression is one of fear. (After Bolwig, N., *Behaviour*, **14**, 136–65, 1959.)

discussed in Chapter 4. There can be no doubt that the dominance hier-archy is based upon the threat of physical violence although, as we have just mentioned, it is not always employed. Primates vary greatly in the degree of fighting that is observed both within and between groups. Even within a species there are considerable variations; langur populations in northern India are far more peaceful than those in the south. The popula-tion density is far greater in the area where the aggressive groups live and this is probably one factor involved. Fighting is very common in crowded zoo colonies, which rarely have sufficient space for subordinate animals to keep out of the way of more dominant ones. Russell and Russell[225A] lay great emphasis on the effects of density upon aggression in both human and non-human societies. It clearly *is* a critical factor under certain circumstances, but as we shall see when discussing the social organization of two different baboons (see p. 266), quite irrespective of density, some types of social structure depend more than others on overt aggression for their mainten-ance and stability.

The nature of the ranking system is such that its age and sex determine to a great extent the rôle a primate plays in its group. One recalls McBride's use of the term 'caste' to apply to an analogous situation in the fowl. However there are, as might be expected, further complications in the primate societies. For example, the Japanese scientists working on *Macaca fuscata* and those working with free-ranging groups of the rhesus monkey (*M. mulatta*) have found that young monkeys tend to 'inherit' the rank of their mother. Thus sons of mothers who rank high have a better chance than most of gaining rank amongst the males (see Sade[229A]).

Sometimes the rank of an animal seems to effect not just its rôle in the group, but the actual position it takes up on the ground. Figs. 10.9 and 10.10 illustrate two examples of this. The former shows a group of baboons (*Papio cynocephalus*) on the march. The most dominant males are in the centre, close to the females with small infants. Two males are in close con-sort relationships with females in oestrus, whilst less dominant adult males and young animals move peripherally. Fig. 10.10 illustrates more diagrammatically the structure of a troop of the Japanese macaque. Again the dominant male 'leaders' are in the centre with females and their young. As males grow up they are driven to the periphery. The arrows at the top indicate the movements of solitary males, not uncommon in the Japanese macaque who are not attached to any troop but, when they do make contact, avoid the leaders. Both in baboons and macaques leaders tend to support one another in asserting their dominance. As males become mature they begin to challenge the leaders and individual groups have now been kept under observation for long enough to record new males taking up dominant positions and gaining access to females.

It is worth noting here that it is commonly at this stage, when young

Fig. 10.9 A troop of baboons on the march through open country. The adult males are distinguished by their large size and well developed manes. Females with babies more in the centre. The two females in oestrus (shown by dark hind parts) move in consortship with the most dominant males. Further explanation in the text. (From Hall and De Vore,[936] 1965, in *Primate Behaviour: Field Studies of Monkeys and Apes*. Holt, Rinehart and Winston, New York and London.)

adult males are challenging for positions of dominance, that some social disruption occurs and males may leave their group. They may wander alone for a time but many subsequently become attached to another group. Lindburg[153B] has described a quite frequent changing between groups in wild male rhesus macaques, most commonly during the main breeding

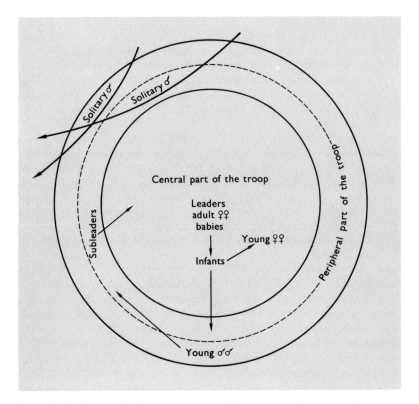

Fig. 10.10 Diagrammatic representation of the structure of a troop of Japanese monkeys. The central part is occupied by dominant males and females with babies. Arrows indicate how young males are driven to the periphery, but challenge the leaders as they grow up. (From Imanishi, K. In Southwick,[248A] *Primate Social Behavior*. Van Nostrand. Princeton.)

season. Rowell[224A] has also seen interchange of adult males between groups of baboons which she could not relate to dominance disputes. Such changes are of extreme importance biologically, because they will prevent groups becoming too inbred. The very stability of their social structure might otherwise threaten their fitness in this respect.

Not all primate workers have found such regular spacing as we described above in the baboons and the Japanese monkeys. In part such findings may be due to special conditions. For the baboons living in open country with the constant threat of predation, this kind of spacing ensures good protection for the females and young. Hall and De Vore[93C] describe how at any threat to the troop the largest males rush to investigate. The baboon troops studied by Rowell[224A] did not show nearly such regular spacing, although large males tended to lead and bring up the rear as the troop moved. Her troops were living in a more forested habitat and she found that they invariably reacted to danger by avoidance rather than defence.

The Japanese monkey's regular spatial organization shown in Fig. 10.10, may well be an artifact of their particular situation. The detailed work on these monkeys has been greatly helped by the success of the Japanese workers in getting them to visit artificial stations where food is regularly provided (see Frisch[76A]). Some troops stay close to these stations and providing food in one spot is likely to reinforce dominance structure. Goodall[84] found just the same thing when she provided caches of bananas for her chimpanzees. The food acts as a focus which all animals are trying to approach, the dominant males sit at the best places and the rest of the group tends to space out according to rank. A less rigid system probably prevails when food has to be searched for individually.

A dominance hierarchy in a group with several adult males, although common, is not the only type of organization. Some primate societies are based on 'one-male groups' which consist of a single adult male, several females and their offspring. This is bound to leave a surplus of adult males and these usually form all-male bands whose members occasionally challenge those with harems. This type of organization is found in the plains-living patas monkey, the gelada and the hamadryas baboon (*Papio hamadryas*). In the latter two, numbers of one-male groups are associated together, particularly in the gelada where hundreds of animals move around together in a loose herd structure.[55B] The hamadryas baboon lives in semi-desert regions of north east Africa, and the groups are fairly independent during the day, spreading out to comb the area for food. They come together as a troop at night to sleep on cliffs which provide protection from predators.

The hamadryas baboon has been the subject of some remarkable field experiments by Kummer[142C, 142D] which provide an excellent demonstration of the flexibility of primate behaviour. In the one-male groups, or 'units' as Kummer calls them, the male closely 'herds' his 3 or 4 females. They move with him at all times and if as they forage one strays too far from him, the male first threatens her and, if she does not return to his side, chases and bites her. Customarily females return to the male at once when he 'stares' at them—a low intensity threat gesture. The yellow baboon,

Papio cynocephalus is a close relative, and in fact hybridizes with *P. hamadryas*. This is the baboon studied by Hall and De Vore, whose troop structure was described above. Females in *cynocephalus* troops do not associate with particular males save at their brief oestrus period when they form a temporary consort relationship with a dominant male. For most of the time they move freely and associate with many different members of the troop.

Kummer transplanted a few *cynocephalus* females into a troop of *hamadryas*. Very quickly males began to herd them into their own units. At first the females wandered away and, if they responded at all to a male's threat, it was to flee. This of course produced the opposite effect to that which they intended. The male instantly pursued them and drove them back to his harem. Kummer found that within a few hours the *cynocephalus* females had learnt their harsh lesson and stayed close to their male.

This result forces us to recognize as a real possibility that some of the variation in social organization which we observe between primate species is of cultural and not genetic origin. The young *hamadryas* baboon grows up in a one male unit, the young *cynocephalus* in the less restricted social climate of the larger group. They may have similar potential and yet develop quite differently. In Chapter 7 we discussed the cultural transmission of minor items of behaviour such as food washing habits, but clearly with the long infancy of primates cultural effects can extend much beyond this.

In fact we have already mentioned considerable variations in behaviour even between different groups of the same species, as with the baboons of open habitats as compared with those from more forested areas, (see p. 266). In the langur (*Presbytis entellus*) of India there are conspicuous differences between the organization of some groups in the south and more northerly populations. In the former one-male groups are common with attendant all-male groups, whereas to the north multi-male groups are the rule (see Yoshiba[287B]).

How far could such differences be adaptive? The southern populations are much more densely packed into the available habitat, and natural predators have largely been removed by human influence. In addition the climate is more severe than in the north, particularly in summer when there are long periods of drought. Is population density the basic factor behind all the behavioural differences, or does social organization provide more specific adaptations to environmental circumstances? It will require long term studies to provide answers to many of the obvious questions.

Finally we may mention one other aspect of primate inter-relationships which is as good an example as any of the range of variability which they provide. It has frequently been recognized that the infants in a group represent a powerful focus of attention. Many individuals other than its

mother approach an infant, attempting to groom and play with it. The degree to which the mother allows such contacts is surprisingly variable. Langur mothers are very permissive and allow other females to take their babies from time to time; sometimes even females from other troops when they intermingle. Rhesus macaque mothers scarcely allow the baby to leave them for several weeks and thereafter restrict its contact with other individuals.

A close relative of the rhesus, the Barbary macaque behaves quite differently. Mothers allow other individuals take their babies, even within a few days of birth. Young babies can be seen moving about in the group not obviously under the care of any particular adult. Males as well as females seek contact with babies and Deag and Crook[61] describe a most remarkable relationship between males and infants. The start of a sequence is illustrated in Plate 5; a young male encourages an infant to approach and ride on his back. The male will then take the baby up to another male, often a high-ranking adult and 'present' it. The exact function of this behaviour is obscure at present, but its effect is to allow a subordinate animal to stay close to a dominant where, without the baby as a 'buffer', it would not be permitted.

It is interesting to speculate about the effect upon the infants development of the permissive type of regime in the Barbary macaque as compared to the more restricted maternal care of the rhesus. Analogies with diverse human child-rearing practices spring to mind—perhaps too readily.

Certainly, quite apart from any light they may throw on the origins of human social life (see reviews by Jay,[131A] and Washburn and De Vore[273A]) the primates offer marvellous material for the study of behavioural development. The experimental approach, which must include cross-fostering infants and cultural transplantation, has scarcely begun. If we allow the primates room to live in the natural habitats to which they have adapted, a combination of laboratory studies and field work is certain to yield fascinating and important results.

Plate 5   A subadult male Barbary macaque and a baby (the same animals were seen in Plate 4) in a typical encounter. The two are sitting close to each other (top), the male turns towards the baby and makes the 'teeth-chatter' face—a friendly expression—and the baby approaches him. The baby leaps onto his back, to be carried off and perhaps 'presented' to a higher ranking male. Further details of such behaviour are given above. (Photos by John Deag.)

References

1. ALEXANDER, R. D. (1960). Sound communication in Orthoptera and Cicadidae. In *Animal Sounds and Communication*, ed. LANYON, W. E. and TAVOLGA, W. N., pp. 38–92. Publ. No. 7, Amer. Inst. Biol. Sci., Washington D.C.
2. ALEXANDER, R. D. (1962). Evolutionary change in cricket acoustical communication. *Evolution*, **16**, 443–67.
3. ALEXANDER, R. D. (1962). The role of behavioral study in cricket classification. *Syst. Zool.*, **11**, 53–72.
3A. ALEXANDER, R. D. and BIGELOW, R. S. (1960). Allochronic speciation in field crickets and a new species, *Acheta veletis*. *Evolution*, **14**, 334–46.
4. ALLEE, W. C. (1938). *The Social Life of Animals*. Norton, New York.
4A. ALTMANN, S. A., ed. (1967). *Social Communication among Primates*. University of Chicago Press, Chicago and London.
5. ANDERSSON, B. (1953). The effect of injections of hypertonic NaCl solutions into different parts of the hypothalamus of goats. *Acta physiol. scand.*, **28**, 188–201.
6. ANDERSSON, B. and MCCANN, S. M. (1955). A further study of polydipsia evoked by hypothalamic stimulation in the goat. *Acta physiol. scand.*, **33**, 333–46.
6A. ANDERSSON, B. and MCCANN, S. M. (1955). Drinking, antidiuresis and milk ejection from electrical stimulation within the hypothalamus of the goat. *Acta physiol. scand.*, **35**, 191–201.
7. ANDJUS, R. K., KNÖPFELMACHER, F., RUSSELL, R. W. and SMITH, A. U. (1955) Effects of hypothermia on behaviour. *Nature, Lond.*, **176**, 1015–16.
8. ANDREW, R. J. (1956). Some remarks on behaviour in conflict situations, with special reference to *Emberiza* spp. *Br. J. Anim. Behav.*, **4**, 41–5.
9. ANDREW, R. J. (1956). Intention movements of flight in certain passerines and their use in systematics. *Behaviour*, **10**, 179–204.
9A. ANDREW, R. J. (1969). The effects of testosterone on avian vocalizations. In *Bird Vocalizations*, pp. 97–130 (see Ref. 112A).
9B. ARDREY, R. (1966). *The Territorial Imperative*. Atheneum, New York.
10. BAERENDS, G. P. (1959). The ethological analysis of incubation behaviour. *Ibis*, **101**, 357–68.
11. BAERENDS, G. P. and BAERENDS-VAN ROON, J. M. (1950). An introduction to the study of the ethology of cichlid fishes. *Behaviour Suppl.*, **1**, 1–243.
11A. BAGGERMAN, B. (1966). On the endocrine control of reproductive behaviour in the male three-spined stickleback (*Gasterosteus aculeatus* L.). *Symp. Soc. Exp. Biol.*, **20**, 427–56.
12. BARLOW, H. B. (1961). The coding of sensory messages. In *Current Problems in Animal Behaviour*, pp. 331–60 (see Ref. 257).
13. BARNETT, S. A. (1958). Exploratory behaviour. *Br. J. Psychol.*, **49**, 289–310.
14. BARNETT, S. A. (1963). *A Study in Behaviour*. Methuen, London.
15. BARNETT, S. A. (1964). The concept of stress. *Viewpoints in Biology*, **3**, 170–218. Butterworth, London.

16. BARNETT, S. A. and BURN, J. (1967). Early stimulation and maternal behaviour. *Nature, Lond.*, **213**, 150–52.
16A. BARTH, R. H. (1968). The comparative physiology of reproductive processes in cockroaches. Part I. Mating behaviour and its endocrine control. *Adv. Reprod. Physiol.*, **3**, 167–207.
17. BASTOCK, M. (1956). A gene mutation which changes a behavior pattern *Evolution*, **10**, 421–39.
18. BASTOCK, M. (1967). *Courtship: a Zoological Study*. Heinemann Educational, London.
19. BASTOCK, M. and MANNING, A. (1955). The courtship of *Drosophila melanogaster. Behaviour*, **8**, 85–111.
20. BATESON, P. P. G. (1964). Effect of similarity between rearing and testing conditions on chicks' following and avoidance responses. *J. comp. physiol. Psychol.*, **57**, 100–3.
21. BATESON, P. P. G. (1966). The characteristics and context of imprinting. *Biol. Rev.*, **41**, 177–220.
21A. BATESON, P. P. G. and REESE, E. P. (1968). Reinforcing properties of conspicuous objects before imprinting has occurred. *Psychon. Sci.*, **10**, 379–80.
21B. BATESON, P. P. G. and REESE, E. P. (1969). The reinforcing properties of conspicuous stimuli in the imprinting situation. *Anim. Behav.*, **17**, 692–99.
22. BEACH, F. A. (1942). Analysis of the stimuli adequate to elicit mating behavior in the sexually inexperienced male rat. *J. comp. Psychol.*, **33**, 163–207.
23. BEACH, F. A. (1942). Male and female mating behavior in prepuberally castrated female rats treated with androgens. *Endocrinology*, **31**, 673–78.
23A. BEACH, F. A. (1950). The Snark was a Boojum. *Am. Psychol.*, **5**, 115–24.
24. BEACH, F. A. (1961). *Hormones and Behavior*, 2nd edn. Cooper Square Publishers, New York.
25. BEACH, F. A. and LEVINSON, G. (1950). Effects of androgen on the glans penis and mating behavior of castrated male rats. *J. exp. Zool.*, **114**, 159–71.
26. BELLOWS, R. T. (1939). Time factors in water-drinking in dogs. *Am. J. Physiol.*, **125**, 87–97.
26A. BENNET-CLARK, H. C. (1970). The mechanism and efficiency of sound production in mole crickets. *J. exp. Biol.*, **52**, 619–52.
26B. BENNET-CLARK, H. C. (1970). A new French mole cricket, differing in song and morphology from *Gryllotalpa gryllotalpa* L. (Orthoptera: Gryllotalpidae). *Proc. R. ent. Soc. Lond.* (B), 39, 125–32.
27. BITTERMAN, E. R. (1965). The evolution of intelligence. *Scient. Am.*, **212** (1), 92–100.
28. BLAIR, W. F. (1955). Mating call and stage of speciation in the *Microhyla olivacea—M. carolinensis* complex. *Evolution*, **9**, 469–80.
29. BLAIR, W. F. (1958). Mating call in the speciation of anuran amphibians. *Am. Nat.*, **92**, 27–51.
30. BLEST, A. D. (1957). The function of eyespot patterns in the Lepidoptera. *Behaviour*, **11**, 209–56.
31. BLEST, A. D. (1957). The evolution of protective displays in the Saturnioidea and Sphingidae (Lepidoptera). *Behaviour*, **11**, 257–309.
32. BLEST, A. D. (1961). The concept of ritualization. In *Current Problems in Animal Behaviour*, pp. 102–24 (see Ref. 257).

32A. BLOMFIELD, S. and MARR, D. (1970). How the cerebellum may be used. *Nature, Lond.*, **227**, 1224–28.

32B. BLURTON-JONES, N. G. (1959). Experiments on the causation of the threat postures of Canada geese. *Rep. Severn Wildfowl Trust*, 1960, 46–52.

32C. BLURTON-JONES, N. G. (1968). Observations and experiments on causation of threat displays of the great tit (*Parus major*). *Anim. Behav. Monogr.*, **1**, 75–158.

33. BOWLBY, J. (1953). *Child Care and the Growth of Love*. Penguin, Harmondsworth.

34. BOYD, H. and FABRICIUS, E. (1965). Observations on the incidence of following of visual and auditory stimuli in naïve mallard ducklings (*Anas platyrhynchos*). *Behaviour*, **25**, 1–15.

35. BRADDOCK, J. C. and BRADDOCK, Z. I. (1959). Development of nesting behaviour in the Siamese fighting fish *Betta splendens*. *Anim. Behav.*, **7**, 222–32.

35A. BRELAND, K. and BRELAND, M. (1961). The misbehavior of organisms. *Am. Psychol.*, **16**, 681–84.

36. BROADBENT, D. E. (1961). *Behaviour*. Eyre and Spottiswoode, London.

37. BROADHURST, P. L. (1969). Psychogenetics of emotionality in the rat. *Ann. N. Y. Acad. Sci.*, **159**, 806–24.

38. BROADHURST, P. L. and LEVINE, S. (1963). Behavioural consistency in strains of rats selectively bred for emotional elimination. *Br. J. Psychol.*, **54**, 121–25.

38A. BRONSON, F. H. and DESJARDINS, C. (1968). Aggression in adult mice: modification by neonatal injections of gonadal hormones. *Science, N. Y.*, **161**, 705–6.

39. BROWN, J. L. and HUNSPERGER, R. W. (1963). Neuroethology and the motivation of agonistic behaviour. *Anim. Behav.*, **11**, 439–48.

39A. BRUNER, J. and KENNEDY, D. (1970). Habituation: occurrence at a neuromuscular junction. *Science, N. Y.*, **169**, 92–94.

40. BUCKLEY, P. A. (1969). Disruption of species-typical behaviour patterns in F_1 hybrid *Agapornis* parrots. *Z. Tierpsychol.*, **26**, 737–43.

41. BULL, H. O. (1957). Conditioned responses. In *The Physiology of Fishes*, 2 vols. ed. BROWN, M. E., pp. 211–28. Academic Press, New York.

42. BULLOCK, T. (1961). The origins of patterned nervous discharge. *Behaviour*, **17**, 48–59.

42A. BUTLER, C. G. (1954). *The World of the Honeybee*. Collins, London.

43. CARMICHAEL, L. (1926). The development of behavior in vertebrates experimentally removed from the influence of external stimulation. *Psychol. Rev.*, **33**, 51–58.

44. CARMICHAEL, L. (1927). A further study of the development of behavior in vertebrates experimentally removed from the influence of external stimulation. *Psychol. Rev.*, **34**, 34–47.

44A. CARPENTER, C. R. (1965). The howlers of Barro Colorado island. In *Primate Behavior*, pp. 250–91 (see Ref. 66A).

44B. CARTER, C. S. and MARR, J. N. (1970). Olfactory imprinting and age variables in the guinea-pig, *Cavia porcellus*. *Anim. Behav.*, **18**, 238–44.

45. CARTHY, J. D. (1958). *An Introduction to the Behaviour of Invertebrates*. Allen and Unwin, London.

46. CARTHY, J. D. (1965). *The Behaviour of Arthropods*. Oliver and Boyd, Edinburgh.

47. CARTHY, J. D. and EBLING, F. J., eds. (1964). *The Natural History of Aggression*. Academic Press, London and New York.
48. CHANCE, E. P. (1940). *The Truth about the Cuckoo*. Country Life, London.
49. CHANCE, M. R. A. (1962). An interpretation of some agonistic postures: the role of 'cut-off' acts and postures. *Symp. zool. Soc. Lond.*, **8**, 71–89.
49A. CHERKIN, A. (1969). Kinetics of memory consolidation: role of amnesic treatment parameters. *Proc. natn. Acad. Sci. U.S.A.*, **63**, 1094–1101.
50. CHITTY, D. (1960). Population processes in the vole and their relevance to general theory. *Can. J. Zool.*, **38**, 99–113.
51. CLARK, E. ARONSON, L. and GORDON, M. (1954). Mating behavior patterns in two sympatric species of xiphophorin fishes: their inheritance and significance in sexual isolation. *Bull. Am. Mus. nat. Hist.*, **103**, 141–225.
51A. CLARK, R. B. (1960). Habituation of the polychaete *Nereis* to sudden stimuli. 1. General properties of the habituation process. *Anim. Behav.* **8**, 82–91.
51B. CLARK, R. B. (1960). Habituation of the polychaete *Nereis* to sudden stimuli. 2. Biological significance of habituation. *Anim. Behav.*, **8**, 92–103.
52. CLARK, R. B. (1965). The learning abilities of Nereid polychaetes and the role of the supra-oesophageal ganglion. *Anim. Behav. Suppl.*, **1**, 89–100.
53. CLARK, W. E. LE GROS (1965). *History of the Primates*, 9th edn. Brit. Mus. (Nat. Hist.), London.
54. CLARKE, J. R. (1955). Influence of numbers on reproduction and survival in two experimental vole populations. *Proc. R. Soc. B.*, **144**, 68–85.
54A. CLAYTON, F. L. and HINDE, R. A. (1968). The habituation and recovery of aggressive display in *Betta splendens*. *Behaviour*, **30**, 96–106.
54B. COONS, E. E., LEVAK, M. and MILLER, N. E. (1965). Lateral hypothalamus: learning of food-seeking response motivated by electrical stimulation. *Science, N.Y.*, **150**, 1320–21.
54C. COULSON, J. C. (1966). The influence of the pair-bond and age on the breeding biology of the kittiwake gull *Rissa tridactyla*. *J. Anim. Ecol.*, **35**, 269–79.
54D. COULSON, J. C. (1968). Differences in the quality of birds nesting in the centre and on the edges of a colony. *Nature, Lond.*, **217**, 478–79.
55. CRANE, J. (1957). Basic patterns of display in fiddler crabs (*Ocypodidae*, genus *Uca*). *Zoologica*, **42**, 69–82.
55A. CROOK, J. H. (1965). The adaptive significance of avian social organizations. *Symp. zool. Soc. Lond.*, **14**, 181–218.
55B. CROOK, J. H. (1966). Gelada baboon herd structure and movement: a comparative report. *Symp. zool. Soc. Lond.*, **18**, 237–58.
55C. CROOK, J. H. (1970). Social organization and the environment: aspects of contemporary social ethology. *Anim. Behav.*, **18**, 197–209.
55D. CROOK, J. H. (1970). The socio-ecology of primates. In *Social Behaviour in Birds and Mammals*, ed. CROOK, J. H., pp. 103–66. Academic Press, London and New York.
56. CRUZE, W. W. (1935). Maturation and learning in chicks. *J. comp. Psychol.*, **19**, 371–409.
57. CULLEN, E. (1957). Adaptations in the kittiwake to cliff-nesting. *Ibis*, **99**, 275–302.

57A. CULLEN, E. (1960). Experiment on the effect of social isolation on reproductive behaviour in the three-spined stickleback. *Anim. Behav.*, **8**, 235.

58. CULLEN, J. M. (1962). The pecking response of young wideawake terns, *Sterna fuscata. Ibis.* **103**, 162–70.

59. CULLEN, J. M. and ASHMOLE, N. P. (1963). The Black Noddy (*Anous tenuirostris*) on Ascension Island. Part 2. Behaviour. *Ibis*, **103**, 423–46.

60. DAANJE, A. (1950). On locomotory movements in birds and the intention movements derived from them. *Behaviour*, **3**, 48–98.

60A. DANE, B. and KLOOT, W. G. van der (1964). An analysis of the display of the goldeneye duck (*Bucephala clangula* L.). *Behaviour*, **22**, 283–328.

60B. DARLING, F. F. (1935). *A Herd of Red Deer.* Oxford University Press, London.

60C. DARLING, F. F. (1938). *Bird Flocks and the Breeding Cycle.* Cambridge University Press, Cambridge.

60D. DAVIS, D. E. (1958). The role of density in aggressive behaviour of house mice. *Anim. Behav.*, **6**, 207–10.

60E. DAWSON, R. G. and MCGAUGH, J. L. (1969). Electroconvulsive shock effects on a reactivated memory trace: further examination. *Science, N.Y.*, **166**, 525–27.

61. DEAG, J. M. and CROOK, J. H. (1971). Social behaviour and 'agonistic buffering' in the wild Barbary macaque *Macaca sylvana* L. *Folia Primat.* **15**, 183–200.

61A. DEAUX, E. and KAKOLEWSKI, J. W. (1970). Emotionally induced increases in effective osmotic pressure and subsequent thirst. *Science, N.Y.*, **169**, 1226–28.

61B. DENENBERG, V. H. (1964). Critical periods, stimulus input, and emotional reactivity: a theory of infantile stimulation. *Psychol. Rev.*, **71**, 335–351.

61C. DENENBERG, V. H. and ROSENBERG, K. M. (1967). Nongenetic transmission of information. *Nature, Lond.*, **216**, 549–60.

61D. DENNY, M. R. and RATNER, S. C. (1970). *Comparative Psychology. Research in Animal Behavior.* Revised Edition. The Dorsey Press. Homewood, Illinois.

62. DETHIER, V. G. (1953). Summation and inhibition following contralateral stimulation of the tarsal chemoreceptors of the blowfly. *Biol. Bull.*, **105**, 257–68.

63. DETHIER, V. G. (1964). Microscopic brains. *Science, N.Y.*, **143**, 1138–45.

64. DETHIER, V. G. and STELLAR, E. (1970). *Animal Behavior*, 3rd edn. Prentice-Hall, Englewood Cliffs, New Jersey.

65. DEUTSCH, J. A. (1960). *The Structural Basis of Behavior.* Cambridge University Press, Cambridge.

66. DEUTSCH, J. A. (1969). The physiological basis of memory. *Ann. Rev. Psychol.*, **20**, 85–104.

66A. DE VORE, I., ed. (1965). *Primate Behavior. Field Studies of Monkeys and Apes.* Holt, Rinehart and Winston, New York and London.

67. DILGER, W. C. (1962). The behavior of lovebirds. *Scient. Am.*, **206**, 88–98.

67A. DIMOND, S. J. (1970). *The Social Behaviour of Animals.* Batsford, London.

68. EATON, R. L. (1970). Group interactions, spacing and territoriality in cheetahs. *Z. Tierpsychol.*, **27**, 481–91.

69. ECCLES, J. C. (1961). The effects of use and disuse on synaptic function.

In *Brain Mechanisms and Learning*, ed. DELAFRESNAYE, J. F., pp. 335–52. Blackwell Scientific, Oxford.

69A. EIBL-EIBESFELDT, I. (1970). *Ethology. The Biology of Behavior*. Holt, Rinehart and Winston, New York.

69B ELLIS, P. E., ed. (1965). Social organization of animal communities. *Symp. zool. Soc. Lond.*, **14**.

69C. ENGLEMANN, F. (1970). *The Physiology of Insect Reproduction*. Pergamon Press, London and New York.

69D. ERICKSON, C. J., BRUDER, R. H., KOMISARUK, B. R. and LEHRMAN, D. S. (1967). Selective inhibition by progesterone of androgen-induced behavior in male ring doves (*Streptopelia risoria*). *Endocrinology*, **81**, 39–44.

69E. ERICKSON, J. G. (1967). Social hierarchy, territoriality, and stress reactions in sunfish. *Physiol. Zool.*, **40**, 40–48.

70. ESTES, R. D. (1969). Territorial behavior of the wildebeeste (*Connochaetes taurinus* Burchell, 1823). *Z. Tierpsychol.*, **26**, 284–370.

70A. ETKIN, W., ed. (1964). *Social Behavior and Organization among Vertebrates*. University of Chicago Press, Chicago and London.

70B. EVANS, S. M. (1968). *Studies in Invertebrate Behaviour*. Heinemann Educational, London.

71. EWER, R. F. (1968). *Ethology of Mammals*. Logos Press Limited, London.

71A. EWING, A. W. (1969). The genetic basis of sound production in *Drosophila pseudoobscura* and *D. persimilis. Anim. Behav.*, **17**, 555–60.

71B. EWING, A. W. and BENNET-CLARK, H. C. (1968). The courtship songs of *Drosophila. Behaviour*, **31**, 288–301.

72. FISHER, A. E. (1964). Chemical stimulation of the brain. *Scient. Am.*, **210**, 60–68.

73. FLEXNER, J. B., FLEXNER, L. B. and STELLAR, E. (1963). Memory in mice as affected by intracerebral puromycin. *Science, N.Y.*, **141**, 57–59.

74. FRAENKEL, G. and GUNN, D. L. (1964). *The Orientation of Animals*. Dover Publications, London.

75. FRANK, B., STEIN, D. G. and ROSEN, J. (1970). Interanimal "memory" transfer: results from brain and liver homogenates. *Science, N.Y.*, **169**, 399–402.

76. FRANZISKET, L. (1953). Untersuchungen zur Spezifität und Kumulierung der Erregungsfähigkeit und zur Wirkung einer Ermüdung in der Afferenz bei Wischbewegung des Rückenmarkfrosches. *Z. vergl. Physiol.*, **34**, 525–38.

76A. FRISCH, J. E. (1968). Individual behavior and intertroop variability in Japanese macaques. In *Primates: Studies in Adaptation and Variability*, pp. 243–52. (See Ref. 131B).

77. FRISCH, K. von, (1967). *The Dance Language and Orientation of Bees*. The Belknap Press of Harvard University Press, Cambridge Mass.

78. FULLER, J. L. and THOMPSON, W. R. (1960). *Behavior Genetics*. Wiley, New York and London.

80. GALAMBOS, R. (1961). Changing concepts of the learning mechanism. In *Brain Mechanisms and Learning*, ed. DELAFRESNAYE, J. F., pp. 231–41. Blackwell Scientific, Oxford.

80A. GARCIA, J. and KOELLING, R. (1966). Relation of cue to consequence in avoidance learning. *Psychonom. Sci.*, **4**, 123–24.

81. GARDNER, B. T. (1964). Hunger and sequential responses in the hunting behavior of salticid spiders. *J. comp. physiol. Psychol.*, **58**, 167–73.

81A. GARDNER, R. A. and GARDNER, B. T. (1969). Teaching sign language to a chimpanzee. *Science, N.Y.*, **165**, 664–72.

81B. GAZZANIGA, M. S. (1967). The split brain in man. *Scient. Am.*, **217**, 24–29.

82. GELLERMAN, L. W. (1933). Form discrimination in chimpanzees and two-year-old children: I form (triangularity) *per se. J. genet. Psychol.*, **42**, 3–27.

82A. GILBERT, R. M. and SUTHERLAND, N. S., eds. (1969). *Animal Discrimination Learning*. Academic Press, London and New York.

82B. GLASSMAN, E. (1969). The biochemistry of learning: an evaluation of the role of RNA and protein. *Ann. Rev. Biochem.*, **38**, 605–46.

83. GOLDSTEIN, A. C. (1957). The experimental control of sex behavior in animals. In *Hormones, Brain Function, and Behavior*, ed. HOAGLAND, H., pp. 99–123. Academic Press Inc. New York.

84. GOODALL, J. van LAWICK (1968). The behaviour of free-living chimpanzees in the Gombe Stream reserve. *Anim. Behav. Monogr.*, 1(3), 161–311.

85. GORBMAN, A. and BERN, H. A. (1962). *A Textbook of Comparative Endocrinology*. Wiley, New York and London.

85A. GOTTLIEB, G. (1968). Prenatal behavior of birds. *Q. Rev. Biol.*, **43**, 148–74.

86. GOY, R. W. and JAKWAY, J. S. (1962). Role of inheritance in determination of sexual behavior patterns. In *Roots of Behavior*, ed. BLISS, E. L., pp. 96–112. Harper, New York.

86A. GREEN, M., GREEN, R. and CARR, W. J. (1966). The hawk-goose phenomenon: A replication and an extension. *Psychon. Sci.*, **4**, 185–6.

86B. GREEN, R., CARR, W. J. and GREEN, M. (1968). The hawk-goose phenomenon: Further confirmation and a search for the releaser. *J. Psychol.*, **69**, 271–76.

87. GREGORY, R. L. (1961). The brain as an engineering problem. In *Current Problems in Animal Behaviour*, pp. 307–30 (see Ref. 257).

88. GRIFFIN, D. R. (1958). *Listening in the Dark*. Yale University Press, New Haven.

88A. GROSSMAN, S. P. (1960). Eating or drinking elicited by direct adrenergic or cholinergic stimulation of the hypothalamus. *Science, N.Y.*, **132**, 301–2.

88B. GROSSMAN, S. P. (1967). *A Textbook of Physiological Psychology*. John Wiley & Sons Inc., New York and London.

88C. GROSSMAN, S. P. (1968). The physiological basis of specific and non-specific motivational processes. *Nebr. Symp. Motiv.*, **16**, 1–46.

89. GRUNT, J. A. and YOUNG, W. C. (1952). Differential reactivity of individuals and the response of the male guinea pig to testosterone propionate. *Endocrinology*, **51**, 237–48.

90. GUITON, P. (1959). Socialisation and imprinting in brown leghorn chicks. *Anim. Behav.*, **7**, 26–34.

91. GUITON, P. (1962). The development of sexual responses in the domestic fowl, in relation to the concept of imprinting. *Symp. zool. Soc. Lond.*, **8**, 227–34.

91A. GWADZ, R. W. (1970). Monofactorial inheritance of early sexual receptivity in the mosquito, *Aëdes atropalpus. Anim. Behav.*, **18**, 358–61.

92. HAILMAN, J. P. (1962). Pecking of laughing-gull chicks to models of the parental head. *Auk*, **79**, 89–98.

93. HAILMAN, J. P. (1965). Cliff-nesting adaptations of the Galapagos swallow-tailed gull. *Wilson Bull.*, **77**, 346–62.

93A. HAILMAN, J. P. (1967). The ontogeny of an instinct. The pecking response in chicks of the laughing gull (*Larus atricilla* L.) and related species. *Behaviour Suppl.*, **15**, 1–159.

93B. HALE, E. B. (1966). Visual stimuli and reproductive behavior in bulls. *J. Anim. Sci.*, **25** (*Suppl.*), 36–44.

93C. HALL, K. R. L. and DE VORE, I. (1965). Baboon social behavior. In *Primate Behavior. Field Studies of Monkeys and Apes*, pp. 53–110. (see Ref. 66A).

94. HALL-CRAGGS, J. (1962). The development of song in the blackbird (*Turdus merula*). *Ibis*, **104**, 277–300.

94A. HALLIDAY, M. S. (1966). Effect of previous exploratory activity on the exploration of a simple maze. *Nature, Lond.*, **209**, 432–33.

95. HARLOW, H. F. (1949). The formation of learning sets. *Psychol. Rev.*, **56**, 51–65.

96. HARLOW, H. F. (1958). The evolution of learning. In *Behavior and Evolution*, ed. ROE, A. and SIMPSON, G. G., pp. 269–90. Yale University Press, New Haven.

97. HARLOW, H. F. and HARLOW, M. K. (1965). The affectional systems. In *Behavior of Nonhuman Primates*, ed. SCHRIER, A. M., HARLOW, H. F. and STOLLNITZ, F., pp. 287–321. Academic Press, New York and London.

98. HARRIS, G. W. and MICHAEL, R. P. (1964). The activation of sexual behaviour by hypothalamic implants of oestrogen. *J. Physiol. Lond.*, **171**, 275–301.

100. HEBB. D. O. (1958). Alice in Wonderland or psychology among the biological sciences. In *Biological and Biochemical Bases of Behavior*, ed. HARLOW, H. F. and WOOLSEY, C. N., pp. 451–67. University of Wisconsin Press, Madison.

101. HEBB, D. O. and WILLIAMS, K. (1946). A method of rating animal intelligence. *J. gen. Psychol.*, **34**, 59–65.

102. HEILIGENBERG, W. (1963). Ursachen für das Auftreten von Instinktsbewegungen bei einem Fische (*Pelmatochromis subocellatus kribensis*). *Z. vergl. Physiol.*, **47**, 339–80.

104. HESS, E. H. (1956). Space perception in the chick. *Scient. Am.*, **195**, 71–80.

105. HESS, E. H. (1964). Imprinting in birds. *Science, N.Y.*, **146**, 1128–39.

107. HINDE, R. (1954). Factors governing the changes in strength of a partially inborn response, as shown by the mobbing behaviour of the chaffinch (*Fringilla coelebs*). I. The nature of the response, and an examination of its course. *Proc. R. Soc. B*, **142**, 306–31.

108. HINDE, R. (1954). Factors governing the changes in strength of a partially inborn response, as shown by the mobbing behaviour of the chaffinch (*Fringilla coelebs*) II. The waning of the response. *Proc. R. Soc. B*, **142**, 331–58.

109. HINDE, R. A. (1958). The nest-building behaviour of domesticated canaries. *Proc. zool. Soc. Lond.*, **131**, 1–48.

110. HINDE, R. A. (1959). Unitary drives. *Anim. Behav.*, **7**, 130–41.

111. HINDE, R. A. (1960). Factors governing the changes in strength of a partially inborn response, as shown by the mobbing behaviour of the chaffinch (*Fringilla coelebs*). III. The interaction of short-term and long-term incremental and decremental effects. *Proc. R. Soc. B*, **153**, 398–420.

112. HINDE, R. A. (1960). Energy models of motivation. *Symp. Soc. exp. Biol.*, **14**, 199–213.

112A. HINDE, R. A., ed. (1969). *Bird Vocalizations*. Cambridge University Press, Cambridge.

112B. HINDE, R. A. (1970). *Animal Behaviour*, 2nd. edn. McGraw-Hill. New York and London.

112C. HINDE, R. A. and FISHER, J. (1952). Further observations on the opening of milk bottles by birds. *Brit. Birds*, **44**, 393–96.

112D. HINDE, R. A. and SPENCER-BOOTH, Y. (1971). Effects of brief separation from mother on rhesus monkeys. *Science, N.Y.*, **173**, 111–18.

113. HINDE, R. A. and STEEL, E. (1964). Effect of exogenous hormones on the tactile sensitivity of the canary brood patch. *J. Endocr.*, **30**, 355–59.

114. HINDE, R. A. and STEEL, E. (1966). Integration of the reproductive behaviour of female canaries. *Symp. Soc. exp. Biol.*, **20**, 401–26.

115. HINDE, R. A. and TINBERGEN, N. (1958). The comparative study of species-specific behavior. In *Behavior and Evolution*, ed. ROE, A. and SIMPSON, G. G., pp. 251–68. Yale University Press, New Haven.

116. HIRSCH, J., ed. (1967). *Behavior-Genetic Analysis*. McGraw-Hill, New York and Maidenhead.

117. HIRSCH, J., LINDLEY, R. H. and TOLMAN, E. C. (1955). An experimental test of an alleged innate sign stimulus. *J. comp. physiol. Psychol.*, **48**, 278–80.

117A. HODOS, W. and CAMPBELL, C. B. G. (1969). *Scala Naturae*: Why there is no theory in comparative psychology. *Psychol. Rev.*, **76**, 337–50.

117B. HÖHN, E. O. (1970). Gonadal hormone concentrations in northern phalaropes in relation to nuptial plumage. *Canad. J. Zool.*, **48**, 400–1.

117C. HOKANSON, J. E. (1969). *The Physiological Bases of Motivation*. John Wiley & Sons. Inc., New York and London.

118. HOLST, E. VON and SAINT-PAUL, U. VON (1963). On the functional organisation of drives. *Anim. Behav.* **11**, 1–20.

118A. HORN, G. and HINDE, R. A. (1970). *Short-term Changes in Neural Activity and Behaviour*. Cambridge University Press, Cambridge.

119 HOYLE, G. (1970). Cellular mechanisms underlying behavior-neuro-ethology. *Adv. Ins. Physiol.*, **7**, 349–444.

120. HUBER, F. (1962). Central nervous control of sound production in crickets and some speculations on its evolution. *Evolution*, **16**, 429–42.

122. HUNSAKER, D. (1962). Ethological isolating mechanisms in the *Sceloporus torquatus* group of lizards. *Evolution*, **16**, 62–74.

123. HUNTER, I. M. L. (1964). *Memory*. Penguin, Harmondsworth.

124. HUTCHISON, J. B. (1970). Influence of gonadal hormones on the hypo-thalamic integration of courtship behaviour in the Barbary dove. *J. Reprod. Fert., Suppl.*, **11**, 15–41.

125. HUXLEY, J. S. (1914). The courtship habits of the great crested grebe (*Podiceps cristatus*). *Proc. zool. Soc. Lond.*, ***1914***(2), 491–562.

128 IERSEL, J. J. A. VAN and BOL, A. C. A. (1958). Preening of two tern species. A study on displacement activities. *Behaviour*, **13**, 1–88.

128A. IKEDA, K. and KAPLAN, W. D. (1970). Patterned neural activity of a mutant *Drosophila melanogaster*. *Proc. natn. Acad. Sci. U.S.A.*, **66**, 765–72.

128B. IKEDA, K. and KAPLAN, W. D. (1970). Unilaterally patterned neural activity of gynandromorphs, mosaic for a neurological mutant of *Drosophila melanogaster*. *Proc. natn. Acad. Sci. U.S.A.*, **67**, 1480–87.

129. IMMELMANN, K. (1969). Über den Einfluss frühkindlicher Erfahrungen auf die geschlechtliche Objektfixierung bei Estrildiden. *Z. Tierpsychol.*, **26**, 677–91.

130. ITÔ, Y. (1970). Groups and family bonds in animals in relation to their habitat. In *Development and Evolution of Behavior*, ed. ARONSON, L. R., TOBACH, E., LEHRMAN, D. S. and ROSENBLATT, J. S. pp. 389–415. W. H. Freeman, San Francisco.

131. JANOWITZ, H. D. and GROSSMAN, M. I. (1949). Some factors affecting the food intake of normal dogs and dogs with esophagotomy and gastric fistulas. *Am. J. Physiol.*, **159**, 143–48.

131A. JAY, P. (1968). Primate field studies and human evolution. In *Primates: Studies in Adaptation and Variability*, pp. 487–503 (see Ref. 131B).

131B. JAY, P. C., ed. (1968). *Primates: Studies in Adaptation and Variability*. Holt, Rinehart and Winston, New York and London.

131C. JOFFE, J. M. (1965). Genotype and prenatal and premating stress inter-act to affect adult behavior in rats. *Science, N.Y.*, **150**, 1844–45.

131D. JOHN, E. R. (1967). *Mechanisms of Memory*. Academic Press. New York and London.

132. JOHNS, J. E. (1964). Testosterone induced nuptial feathers in phalaropes. *Condor*, **66**, 449–55.

132A. JOLLY, A. (1966). *Lemur Behavior. A Madagascar Field Study*. University of Chicago press, Chicago and London.

133. KEAR, J. (1964). Colour preference in young Anatidae. *Ibis*, **106**, 361–69.

134. KEAR, J. (1966). The pecking response of young coots *Fulica atra* and moorhens *Gallinula chloropus*. *Ibis*, **108**, 118–22.

135. KELLOGG, W. N. (1968). Communication and language in the home-raised chimpanzee. *Science, N.Y.*, **162**, 423–27.

136. KENNEDY, J. S. (1958). The experimental analysis of aphid behaviour and its bearing on current theories of instinct. *Proc. 10th Int. Congr. Ent.*, *Montreal*, 1956, **2**, 397–404.

137. KENNEDY, J. S. (1965). Co-ordination of successive activities in an aphid. Reciprocal effects of settling on flight. *J. exp. Biol.*, **43**, 489–509.

138. KLOPFER, P. H. (1959). An analysis of learning in young Anatidae. *Ecology*, **40**, 90–102.

139. KLOPFER, P. H., ADAMS, D. K. and KLOPFER, M. S. (1964). Maternal 'imprinting' in goats. *Proc. natn. Acad. Sci. U.S.A.*, **52**, 911–14.

140. KÖHLER, W. (1927). *The Mentality of Apes*, 2nd edn. Kegan Paul, Trench, Trubner & Co., London.

140A. KOMISARUK, B. R. (1967). Effects of local brain implants of progesterone on reproductive behavior in ring doves. *J. comp. physiol. Psychol.*, **64**, 219–24.

141. KONISHI, M. (1965). The rôle of auditory feedback in the control of vocalization in the white-crowned sparrow. *Z. Tierpsychol.*, **22**, 770–83.

142. KONORSKI, J. (1948). *Conditioned Reflexes and Neuron Organization*. Cambridge University Press, Cambridge.

142A. KRUIJT, J. P. (1964). Ontogeny of social behaviour in Burmese red jungle fowl (*Gallus gallus spadiceus*) Bonaterre. *Behaviour Suppl.*, **12**, 1–201.

142B. KRUIJT, J. P. and HOGAN, J. A. (1967). Social behavior on the lek in black grouse, *Lyrurus tetrix tetrix* (L.). *Ardea*, **55**, 203–40.

142C. KUMMER, H. (1968). Social organization of hamadryas baboons. A field study. *Bibliotheca Primatologica* No. 6, 1–189.

142D. KUMMER, H. (1968). Two variations in the social organization of baboons. In *Primates: Studies in Adaptation and Variability*, pp. 293–312 (see Ref. 131B).

143. LACK, D. (1966). *Population Studies of Birds*. Oxford University Press, London.

145. LASHLEY, K. S. (1950). In search of the engram. *Symp. Soc. exp. Biol.*, **4**, 454–82.

146. LASHLEY, K. S. and WILEY, L. E. (1933). Studies of cerebral function in learning. IX. Mass action in relation to the number of elements in the problem to be learned. *J. comp. Neurol.*, **57**, 3–55.

147. LEHRMAN, D. S. (1953). A critique of Konrad Lorenz's theory of instinctive behavior. *Q. Rev. Biol.*, **28**, 337–63.

147A. LEHRMAN, D. S. (1955). The physiological basis of parental feeding behaviour in the ring dove (*Streptopelia risoria*). *Behaviour*, **7**, 241–86.

148. LEHRMAN, D. S. (1961). Hormonal regulation of parental behavior in birds and infrahuman mammals. In *Sex and Internal Secretions*, 3rd edn., ed. YOUNG, W. C., pp. 1268–1382. Bailliere, Tindall and Cox, London.

149. LEHRMAN, D. S. (1964). The reproductive behavior of ring doves. *Scient. Am.*, **211**, 48–54.

149A. LENNENBERG, E. H. (1967). *Biological Foundation of Language*. Wiley, New York.

150. LETTVIN, J. Y., MATURANA, H. R., MCCULLOCH, W. S. and PITTS, W. H. (1959). What the frog's eye tells the frog's brain. *Proc. Inst. Radio Engrs*, **47**, 1940–51.

150A. LEUTHOLD, W. (1966). Variations in territorial behavior of Uganda kob *Adenota kob thomasi* (Neumann 1896). *Behaviour*, **27**, 215–58.

150B. LEVINE, S. (1966). Sex differences in the brain. *Scient. Amer.*, **214** (4), 84–90.

151. LEVINE, S., HALTMEYER, G. C., KARAS, G. G. and DENENBERG, V. H. (1967). Physiological and behavioral effects of infant stimulation. *Physiol. Behav.*, **2**, 55–59.

151A. LEVINE, S. and MULLINS, R. F. (1966). Hormonal influences on brain organization in infant rats. *Science, N.Y.*, **152**, 1585–92.

152. LEYHAUSEN, P. (1956). Verhaltensstudien an Katzen, *Z. Tierpsychol.*, *Beiheft*, **2**, 1–120.

152A. LEYHAUSEN, P. (1964). The communal organization of solitary mammals. *Symp. zool. Soc. Lond.*, **14**, 249–263.

153. LIND, H. (1959). The activation of an instinct caused by a 'transitional action'. *Behaviour*, **14**, 123–35.

153A. LINDAUER, M. (1961). *Communication among Social Bees*. Harvard University Press, Cambridge, Mass.

153B. LINDBURG, D. G. (1969). Rhesus monkeys: Mating season mobility in adult males. *Science, N.Y.*, **166**, 1176–78.

154. LISSMANN, H. W. (1963). Electric location by fishes. *Scient. Am.* **208**, 50–59.

154A. LOHER, W. and HUBER, F. (1966). Nervous and hormonal control of sexual behaviour in a grasshopper (*Gomphocerus rufus* L.) *Symp. Soc. exp. Biol.*, **20**, 381–400.

154B. LORAINE, J. A. (1970). *Sex and the Population Crisis*. Heinemann Medical Books Ltd., London.

155. LORENZ, K. Z. (1937). The companion in the bird's world. *Auk.*, **54**, 245–73.
156. LORENZ, K. Z. (1941). Vergleichende Bewegungsstudien an Anatinen. *J. Orn., Lpz.*, **89**, 194–293. (An English translation appeared in several parts, in Vols. 57–59 of *Avicultural Magazine*.)
157. LORENZ, K. Z. (1950). The comparative method in studying innate behaviour patterns. *Symp. Soc. exp. Biol.*, **4**, 221–68.
158. LORENZ, K. Z. (1952). *King Solomon's Ring*. Methuen, London. (Reprinted 1959 by Pan, London.)
158A. LORENZ, K. Z. (1958). The evolution of behavior. *Scient. Am.*, **199**(6), 67–78.
159. LORENZ, K. Z. (1966). *Evolution and Modification of Behaviour*. Methuen, London.
160. LORENZ, K. Z. (1966). *On Aggression*. Methuen, London.
161. LORENZ, K. Z. and TINBERGEN, N. (1938). Taxis und Instinkthandlung in der Eirollbewegung der Graugans I. *Z. Tierpsychol.*, **2**, 1–29.
162. LOUCKS, R. B. (1935). The experimental delimitation of neural structures essential for learning: the attempt to condition striped muscle responses with faradization of the sigmoid gyri. *J. Psychol.*, **1**, 5–44.
163. MCBRIDE, G., PARER, I. P. and FOENANDER, F. (1959). The social organization and behaviour of the feral domestic fowl. *Anim. Behav. Monogr.*, **2**, 127–81.
164. MCFARLAND, D. J. (1965). Hunger, thirst and displacement pecking in the Barbary dove. *Anim. Behav.*, **13**, 293–300.
164A. MCFARLAND, D. J. (1966). On the causal and functional significance of displacement activities. *Z. Tierpsychol.*, **23**, 217–35.
165. MCGAUGH, J. L. (1961). Facilitative and disruptive effects of strychnine sulphate on maze learning. *Psychol. Rep.*, **8**, 99–104.
166. MCGAUGH, J. L. (1966). Time-dependent processes in memory storage. *Science, N.Y.*, **153**, 1351–58.
167. MCGAUGH, J. L. and DAWSON, R. G. (1971). Modification of memory storage processes. *Behavl. Sci.*, **16**, 45–63.
168A. MCGILL, T. E. ed. (1965). *Readings in Animal Behavior*. Holt, Rinehart and Winston. New York and London.
168B. MCGILL, T. E. and TUCKER, G. R. (1964). Genotype and sex drive in intact and castrated male mice. *Science, N.Y.*, **145**, 514–15.
169. MACKINTOSH, N. J. (1965). Discrimination learning in the octopus. *Anim. Behav. Suppl.*, **1**, 129–34.
170. MADSEN, M. C. and MCGAUGH, J. L. (1961). The effect on ECS on one-trial avoidance learning. *J. comp. physiol. Psychol.*, **54**, 522–23.
171. MAGNUS, D. B. E. (1958). Experimental analysis of some 'overoptimal' sign stimuli in the mating behaviour of the fritillary butterfly, *Argynnis paphia* L. (Lepidoptera Nymphalidae.) *Proc. 10th Int. Congr. Ent., Montreal*, 1956, **2**, 405–18.
172. MAIER, N. R. F. (1932). Cortical destruction of the posterior part of the brain and its effect on reasoning in rats. *J. comp. Neurol.*, **56**, 179–214.
173. MAIER, N. R. F. and SCHNEIRLA, T. C. (1935). *Principles of Animal Psychology*. McGraw-Hill, New York and Maidenhead.
174. MANNING, A. (1959). The sexual behaviour of two sibling *Drosophila* species. *Behaviour*, **15**, 123–45.
175. MANNING, A. (1961). The effects of artificial selection for mating speed in *Drosophila melanogaster*. *Anim. Behav.*, **9**, 82–92.

176. MANNING, A. (1965). Drosophila and the evolution of behaviour. *Viewpoints in Biology*, **4**, 125–69. Butterworth, London.

177. MANNING, A. (1966). Sexual behaviour. *Symp. R. ent. Soc. Lond.*, **3**, 59–68.

178. MANNING, A. (1966). Corpus allatum and sexual receptivity in female *Drosophila melanogaster*. *Nature, Lond.*, **211**, 1321–22.

178A. MANNING, A. (1967). 'Pre-imaginal conditioning' in *Drosophila*. *Nature, Lond.*, **216**, 338–40.

178B. MANNING, A. (1971). *Evolution of behavior*. In *Psychobiology: Biological Bases of Behavior*, ed. MCGAUGH, J. pp. 1–52. Academic Press, New York and London.

178C. MARLER, P. (1955). Studies of fighting in chaffinches. (1) Behaviour in relation to the social hierarchy. *Brit. J. Anim. Behav.*, **3**, 111–17.

179. MARLER, P. (1959). Developments in the study of animal communication. In *Darwin's Biological Work*, ed. BELL, P. R., pp. 150–206. Cambridge University Press, Cambridge.

179A. MARLER, P. (1969). Colobus guereza: territoriality and group composition. *Science, N.Y.*, **163**, 93–95.

179B. MARLER, P. and HAMILTON, W. J. (1966). *Mechanisms of Animal Behavior*. John Wiley, New York and London.

179C. MARLER, P., KREITH, M. and TAMURA, M. (1964). Song development in hand-raised Oregon juncos. *Auk*, **79**, 12–30.

180. MARLER, P. and TAMURA, M. (1964). Culturally transmitted patterns of vocal behavior in sparrows. *Science, N.Y.*, **146**, 1483–86.

181. MASSERMAN, J. H. (1950). Experimental neuroses. *Scient. Am.*, **182**, 38–43.

182. MAY, D. J. (1949). Studies on a community of willow-warblers. *Ibis*, **91**, 24–54.

183. MAYR, E. (1963). *Animal Species and Evolution*. Oxford University Press, London.

184. MELZACK, R., PENICK, E. and BECKETT, A. (1959). The problem of 'innate fear' of the hawk shape; an experimental study with mallard ducks. *J. comp. physiol. Psychol.*, **52**, 694–98.

185. MICHAEL, R. P. (1966). Action of hormones on the cat brain. In *Brain and Behavior*, ed. GORSKI, R. A. and WHALEN, R. E.; vol. 3, *The Brain and Gonadal Function*, pp. 81–98. University of California Press, Berkeley.

186. MICHAEL, R. P. and SCOTT, P. P. (1964). The activation of sexual behaviour in cats by the subcutaneous administration of oestrogen. *J. Physiol.*, **171**, 254–74.

186A. MILLER, N. E. (1941). The frustration-aggression hypothesis. *Psychol. Rev.*, **48**, 337–42.

187. MILLER, N. E. (1951). Learnable drives and rewards. In *Handbook of Experimental Psychology*, ed. STEVENS, S. S., pp. 435–72. Wiley, New York and London.

188. MILLER, N. E. (1956). Effects of drugs on motivation: the value of using a variety of measures. *Ann. N.Y. Acad. Sci.*, **65**, 318–33.

189. MILLER, N. E. (1957). Experiments on motivation. Studies combining psychological, physiological and pharmacological techniques. *Science N.Y.*, **126**, 1271–78.

190. MILLER, N. E. (1965). Chemical coding of behavior in the brain. *Science, N.Y.*, **148**, 328–38.

191. MILLER, N. E. and KESSEN, M. L. (1952). Reward effects of food via stomach fistula compared with those of food via mouth. *J. comp. physiol. Psychol.*, **45**, 555–64.

192. MOLTZ, H. and STETTNER, L. J. (1961). The influence of patterned-light deprivation on the critical period for imprinting. *J. comp. physiol. Psychol.*, **54**, 279–83.

193. MONTAGU, M. F. A. (1968). Man and Aggression. Oxford University Press. London and New York.

194. MORRIS, D. (1956). The feather postures of birds and the problem of the origin of social signals. *Behaviour*, **9**, 75–113.

195. MORRIS, D. (1957). 'Typical intensity' and its relation to the problem of ritualization. *Behaviour*, **11**, 1–12.

196. MORRIS, D. (1959). The comparative ethology of grassfinches (Erythrurae) and mannikins (Amadinae). *Proc. zool. Soc. Lond.*, **131**, 389–439.

196A. MORRIS, D., ed. (1967). *Primate Ethology*. Weidenfeld and Nicolson, London.

196B. MORRIS, D. (1967). *The Naked Ape*. Jonathan Cape, London.

196C. MOYNIHAN, M. (1955). Some aspects of the reproductive behavior in the black-headed gull (*Larus ridibundus ridibundus* L.) *Behaviour Suppl.*, **4**, 1–201.

196D. MOYNIHAN, M. (1968). Social mimicry; character convergence versus character displacement. *Evolution*, **22**, 315–31.

197. MUNN, N. L. (1950). *Handbook of Psychological Research on the Rat*. Houghton Mifflin, Boston.

198. MYER, J. S. and WHITE, R. T. (1965). Aggressive motivation in the rat. *Anim. Behav.*, **13**, 430–33.

198A. NAPIER, J. R. and NAPIER, P. H., eds. (1970). *Old World Monkeys. Evolution, Systematics and Behavior*. Academic Press. New York and London.

199. NICE, M. M. (1937). Studies in the life history of the song sparrow. I. *Trans. Linn. Soc. N.Y.*, **4**, (Reprinted Dover Publications, New York. T1219, 1964.)

199A. NIESCHLAG, E. and ISMAIL, A. A. A. (1970). Diurnal variations of plasma testosterone in normal and pathological conditions as measured by the technique of competitive protein binding. *J. Endocrin.*, **46**, proc. iii–iv.

199B. NIXON, H. L. and RIBBANDS, C. R. (1952). Food transmission in the honeybee community. *Proc. roy. Soc. B*, **140**, 43–50.

200. NOIROT, E. (1964). Changes in responsiveness to young in the adult mouse. I. The problematical effects of hormones. *Anim. Behav.*, **12**, 52–58.

201. NOIROT, E. (1964). Changes in responsiveness to young in the adult mouse. IV. The effect of an initial contact with a strong stimulus. *Anim. Behav.*, **12**, 442–45.

201A. NOIROT, E. (1969). Changes in responsiveness to young in the adult mouse. V. Priming, *Anim. Behav.* **17**, 542–46.

201B. NORTON-GRIFFITHS, M. (1967). Some ecological aspects of the feeding behaviour of the oystercatcher *Haematopus ostralegus* on the edible mussel. *Mytilus edulis, Ibis*, **109**, 412–24.

201C. NORTON-GRIFFITHS, M. (1969). The organisation, control and development of parental feeding in the oystercatcher. (*Haematopus ostralegus*). *Behaviour*, **34**, 55–114.

202. OLDS, J. (1962). Hypothalamic substrates of reward. *Physiol. Rev.*, **42**, 554–604.

202A. OPPENHEIM, R. W. (1968). Color preference in the pecking response of newly hatched ducks (*Anas platyrhynchos.*) *J. comp. physiol. Psychol., Monogr. Suppl.* (2), 1–17.

202B. OWEN-SMITH, N. (1971). Territoriality in the white rhinoceros (*Ceratotherium simum*) Burchell. *Nature, Lond.*, **231**, 294–96.

203. PAUL, J. (1965). *Cell Biology.* Heinemann Educational, London.

204. PAVLOV, I. P. (1941). *Lectures on Conditioned Reflexes*, 2 vols. International Publishers, New York.

205. PENFIELD, W. (1959). The interpretive cortex. *Science, N.Y.*, **129**, 1719–25.

206. PERDECK, A. C. (1958). The isolating value of specific song in two sibling species of grasshoppers (*Chorthippus brunneus* Thunb. and *C. biguttulus* L.). *Behaviour*, **12**, 1–75.

206A. PETTERSSON, M. (1956). Diffusion of a new habit among green finches. *Nature, Lond.*, **177**, 709–10.

207. PRECHTL, H. F. R. (1953). Zur Physiologie der angeborenen Auslosendenmechanismen. I. Quantitative Untersuchungen über die Sperrbewegung junger Singvögel. *Behaviour*, **5**, 32–50.

207A. PREMACK, D. (1971). Language in chimpanzee? *Science, N.Y.*, **172**, 808–22.

207B. QUARTERMAIN, D., MCEWAN, B. S. and AZMITIA, E. C. (1970). Amnesia produced by electroconvulsive shock or cycloheximide: conditions for recovery. *Science, N.Y.*, **169**, 683–86.

208. RÄBER, H. (1948). Analyse des Balzverhaltens. eines domestizierten Truthahns (*Meleagris*). *Behaviour*, **1**, 237–66.

208A. RALLS, K. (1971). Mammalian scent marking. *Science, N.Y.*, **171**, 443–49.

209. RAMSAY, A. O. (1961). Behaviour of some hybrids in the mallard group. *Anim. Behav.*, **9**, 104–5.

210. RAMSAY, A. O. and HESS, E. H. (1954). A laboratory approach to the study of imprinting. *Wilson Bull.*, **66**, 196–206.

210A. RASA, O. A. E. (1969). The effect of pair isolation on reproductive success in *Etroplus maculatus* (Cichlidae). *Z. Tierpsychol.*, **26**, 846–52.

211. REYNIERS, J. A. (1953). Germ-free life. *Lancet*, 933–34.

212. REYNOLDS, R. W. (1965). An irritative hypothesis concerning the hypothalamic regulation of food intake. *Psychol. Rev.*, **72**, 105–16.

212A. RICE, J. O. and THOMPSON, W. L. (1968). Song development in the indigo bunting. *Anim. Behav.*, **16**, 462–69.

213. RIESEN, A. H. (1947). The development of visual perception in man and chimpanzee. *Science, N.Y.*, **106**, 107–8.

214. RISS, W. (1955). Sex drive, oxygen consumption and heart rate in genetically different strains of male guinea pigs. *Am. J. Physiol.*, **180**, 530–34.

215. ROEDER, K. D. (1967). *Nerve Cells and Insect Behavior*, Revised edn. Harvard University Press, Cambridge, Mass.

216. ROMER, A. S. (1962). *The Vertebrate Body*, 3rd. edn. W. B. Saunders, Philadelphia and London.

219. ROSENBLATT, J. S. and ARONSON, L. R. (1958). The decline of sexual behaviour in male cats after castration with special reference to the role of prior sexual experience. *Behaviour*, **12**, 285–338.

220. ROTH, L. M. (1948). An experimental laboratory study of the sexual behavior of *Aëdes aegypti* (1). *Am. Midl. Nat.*, **40**, 265–352.

221. ROTH, L. M. and BARTH, R. H. (1964). The control of sexual receptivity in female cockroaches. *J. Insect Physiol.*, **10**, 965–75.

222. ROTHENBUHLER, W. C. (1964). Behaviour genetics of nest cleaning in honey-bees. I. Responses of four inbred lines to disease-killed brood. *Anim. Behav.*, **12**, 578–83.

223. ROTHENBUHLER, W. C. (1964). Behavior genetics of nest cleaning in honey-bees. IV. Reponses of F_1 and backcross generations to disease-killed brood. *Am. Zoologist*, **4**, 111–23.

224. ROWELL, C. H. F. (1961). Displacement grooming in the chaffinch. *Anim. Behav.*, **9**, 38–63.

224A. ROWELL, T. (1969). Long-term changes in a population of Ugandan baboons. *Folia primat.*, **11**, 241–54.

225. RUNDQUIST, E. A. (1933). The inheritance of spontaneous activity in rats. *J. comp. Psychol.*, **16**, 415–38.

225A. RUSSELL, C. and RUSSELL, W. M. S. (1968). *Violence, Monkeys and Man.* Macmillan, London.

226. RUSSELL, W. M. S. (1958). Evolutionary concepts in behavioral science; I. Cybernetics, Darwinian theory and behavioral science. *Gen. Syst.* **3**, 18–28.

227. RUSSELL, W. M. S. (1959). Evolutionary concepts in behavioral science; II. Organic evolution and the genetical theory of natural selection. *Gen. Syst.*, **4**, 45–73.

228. RUSSELL, W. M. S. (1961). Evolutionary concepts of behavioral science: III. The evolution of behavior in the individual animal, and the principle of combinatorial selection. *Gen. Syst.*, **6**, 51–91.

229. RUSSELL, W. M. S. (1962). Evolutionary concepts in behavioral science: IV. The analogy between organic and individual behavioral evolution, and the evolution of intelligence. *Gen. Syst.*, **7**, 157–93.

229A. SADE, D. S. (1967). Determinants of dominance in a group of free-ranging rhesus monkeys. In *Social Communication among Primates*, pp. 99–114. (see Ref. 4A).

230. SCHEIN, M. W. and HALE, E. B. (1959). The effect of early social experience on male sexual behaviour of androgen-injected turkeys. *Anim. Behav.*, **7**, 189–200.

231. SCHJELDERUP-EBBE, T. (1935). Social behavior of birds. In *Handbook of Social Psychology*, ed. MURCHISON, C., pp. 947–72. Clark University Press, Worcester Mass.

232. SCHLEIDT, W. M. (1961). Reaktionen von Truthühnern auf fliegende Raubvögel und Versuche zur Analyse ihrer AAM's. *Z. Tierpsychol.*, **18**, 534–60.

232A. SCHLEIDT, W. and SCHLEIDT, M. (1960). Störung der Mutter-Kind-Beziehung bei Truthühnern durch Gehörverlust. *Behaviour*, **16**, 254–60.

233. SCHNIEDER, D. (1966). Chemical sense communication in insects. *Symp. Soc. exp. Biol.*, **20**, 273–97.

233A. SCHNEIRLA, T. C. (1965). Aspects of stimulus and organization in approach/withdrawal processes underlying vertebrate behavioral development. *Adv. Study Behav.*, **1**, 1–71.

234. SCHUTZ, F. (1965). Sexuelle Prägung bei Anatiden. *Z. Tierpsychol.*, **22**, 50–103.

234A. SCOTT, J. P. (1958). *Aggression.* University of Chicago Press, Chicago.

235. SCOTT, J. P. (1962). Critical periods in behavioral development. *Science, N.Y.*, **138**, 949–58.

236. SCOTT, J. P. and FULLER, J. L. (1965). *Genetics and the Social Behavior of the Dog.* University of Chicago Press, Chicago.

236A. SECHZER, J. A. (1970). Prolonged learning and split-brain cats. *Science, N.Y.*, **169**, 889–92.

236B. SELIGMAN, M. E. P. (1970). On the generality of the laws of learning. *Psychol. Rev.* **77**, 406–18.

237. SEVENSTER, P. (1961). A causal analysis of a displacement activity (fanning in *Gasterosteus aculeatus* L.) *Behaviour Suppl.*, **9**, 1–170.

238. SEVENSTER-BOL, A. C. A. (1962). On the causation of drive reduction after a consummatory act (in *Gasterosteus aculeatus* L.). *Archs néerl. Zool.*, **15**, 175–236.

239. SHEPARD, J. F. (1933). Higher processes in the behavior of rats. *Proc. natn. Acad. Sci. U.S.A.*, **19**, 149–52.

240. SHEPPARD, P. M. (1961). Some contributions to population genetics resulting from the study of Lepidoptera. *Adv. Genet.*, **10**, 165–216.

241. SHERRINGTON, C. S. (1906). *The Integrative Action of the Nervous System.* Scribner's, New York.

242. SHERRINGTON, C. S. (1917). Reflexes elicitable in the cat from pinna, vibrissae and jaws. *J. Physiol.*, **51**, 404–31.

243. SHOLL, D. A. (1956). *The Organisation of the Cerebral Cortex.* Methuen, London.

244. SKINNER, B. F. (1938). *The Behavior of Organisms.* Appleton-Century-Crofts, New York.

245. SLUCKIN, W. (1964). *Imprinting and Early Learning.* Methuen, London.

245A. SMITH, J. M. (1964). Group selection and kin selection. *Nature, Lond.*, **201**, 1145–47.

246. SMITH, J. M. (1966). *The Theory of Evolution*, 2nd edn. Penguin, Harmondsworth.

248. SNOW, D. W. (1958). *A Study of Blackbirds.* Allen and Unwin, London.

248A. SOUTHWICK, C. H., ed. (1963). *Primate Social Behavior.* Van Nostrand. Princeton and London.

249. SPALDING, D. (1873). Instinct: with original observations on young animals. *Macmillan's Mag.*, **27**, 282–93. (Reprinted in *Br. J. Anim. Behav.*, **2**, 1–11, 1954.)

250. SPERRY, R. W. (1961). Cerebral organization and behavior. *Science, N.Y.*, **133**, 1749–57.

250A. SPERRY, R. W. (1968). Hemisphere deconnection and unity in conscious awareness. *Amer. Psychol.*, **23**, 723–33.

251. STOKES, A. W. (1962). Agonistic behaviour among blue tits at a winter feeding station. *Behaviour*, **19**, 118–38.

251A. STOUT, J. F. and BRASS, M. E. (1969). Aggressive communication by *Larus glaucescens*. Part II. Visual communication. *Behaviour*, **34**, 42–54.

251B. STOUT, J. F., WILCOX, C. R. and CREITZ, L. E. (1969). Aggressive communication by *Larus glaucescens*. Part 1. Sound communication. *Behaviour*, **34**, 29–41.

251C. SUDD, J. H. (1967). *An Introduction to the Behaviour of Ants.* Edward Arnold, London.

251D. TAYLOR, A., SLUCKIN, W. and HEWITT, R. (1969). Changing colour preference of chicks. *Anim. Behav.*, **17**, 3–8.

252. TEITELBAUM, P. and EPSTEIN, A. N. (1962). The lateral hypothalamic syndrome: recovery of feeding and drinking after lateral hypothalamic lesions. *Psychol. Rev.*, **69**, 74–90.

252A. THIESSEN, D. D. and YAHR, P. (1970). Central control of territorial marking in the Mongolian gerbil. *Physiol. Behav.*, **5**, 275–78.

252B. THOMPSON, T. I. (1963). Visual reinforcement in Siamese fighting fish. *Science, N.Y.*, **141**, 55–57.

252C. THOMPSON, T. I. (1964). Visual reinforcement in fighting cocks. *J. Exptl. Anal. Behav.*, **7**, 45–49.

252D. THOMPSON, W. R., WATSON, J. and CHARLESWORTH, W. R. (1962). The effects of prenatal maternal stress on offspring behavior in rats. *Psychol. Monogr.*, **76**, No. 38.

253. THORPE, W. H. (1950). A note on detour experiments with *Ammophila pubescens* Curt. (Hymenoptera: Sphecidae). *Behaviour*, **2**, 257–63.

254. THORPE, W. H. (1961). *Bird Song*. Cambridge University Press, Cambridge.

255. THORPE, W. H. (1963). *Learning and Instinct in Animals*, 2nd edn. Methuen, London.

256. THORPE, W. H. and DAVENPORT, D., eds. (1965). *Learning and Associated Phenomena in Invertebrates. Anim. Behav. Suppl.*, **1**, 1–190.

257. THORPE, W. H. and ZANGWILL, O. L. eds. (1961). *Current Problems in Animal Behaviour*. Cambridge University Press, Cambridge.

258. TINBERGEN, N. (1951). *The Study of Instinct*. Oxford University Press, London.

259. TINBERGEN, N. (1952). 'Derived' activities; their causation, biological significance, origin and emancipation during evolution. *Q. Rev. Biol.*, **27**, 1–32.

260. TINBERGEN, N. (1953). *The Herring Gull's World*. Collins, London.

261. TINBERGEN, N. (1953). *Social Behaviour in Animals*. Methuen, London.

262. TINBERGEN, N. (1957). On anti-predator responses in certain birds—a reply. *J. comp. physiol. Psychol.*, **50**, 412–14.

263. TINBERGEN, N. (1959). Comparative studies of the behaviour of gulls (Laridae): a progress report. *Behaviour*, **15**, 1–70.

263A. TINBERGEN, N. (1968). On war and peace in men and animals. *Science, N.Y.*, **160**, 1411–18.

263B. TINBERGEN, N., BROEKHUYSEN, G. J., FEEKES, F., HOUGHTON, J. C. W., KRUUK, H. and SZULC, E. (1962). Egg shell removal by the black-headed gull. *Larus ridibundus*, L.; a behaviour component of camouflage. *Behaviour*, **19**, 74–117.

264. TINBERGEN, N., KRUUK, H., PAILETTE, M. and STAMM, R. (1962). How do black-headed gulls distinguish between eggs and egg-shells? *Br. Birds*, **55**, 120–29.

265. TINBERGEN, N. and PERDECK, A. C. (1950). On the stimulus situation releasing the begging response in the newly-hatched herring gull chick (*Larus a. argentatus* Pont.). *Behaviour*, **3**, 1–38.

266. TSCHANZ, B. (1959). Zur Brutbiologie der Trottelume (*Uria aalge aalge* Pont.). *Behaviour*, **14**, 1–100.

266A. VALENSTEIN, E. S., COX, V. C. and KAKOLEWSKI, J. W. (1968). Modification of motivated behavior elicited by electrical stimulation of the hypothalamus. *Science, N.Y.*, **159**, 1119–21.

266B. VALENSTEIN, E. S., COX, V. C. and KAKOLEWSKI, J. W. (1969). Hypothalamic motivational systems: fixed or plastic neural circuits? *Science, N.Y.*, **163**, 1084.

266C. VALENSTEIN, E. S., COX, V. C. and KAKOLEWSKI, J. W. (1970). Re-examination of the role of the hypothalamus in motivation. *Psychol. Rev.*, **77**, 16–31.

267. VALENSTEIN, E. S. and YOUNG, W. C. (1955). An experiential factor influencing the effectiveness of testosterone propionate in eliciting sexual behavior in male guinea pigs. *Endocrinology*, **56**, 173–77.

268. VERNON, J. A. and BUTSCH, J. (1957). Effect of tetraploidy on learning and retention in the salamander. *Science, N.Y.*, **125**, 1033–34.

269. VICTOR, M., ANGEVINE, J. B., MANCALL, E. L. and FISHER, C. M. (1961). Memory loss with lesions of hippocampal formation. *Archs Neurol., Chicago*, **5**, 244–63.

269A. VINCE, M. A. (1969). Embryonic communication, respiration and the synchronization of hatching. In *Bird Vocalizations*, pp. 233–60 (see Ref. 112A).

269B. VOWLES, D. M. (1954). The orientation of ants: I. The substitution of stimuli. *J. exp. Biol.*, **31**, 341–55.

270. VOWLES, D. M. (1965). Maze learning and visual discrimination in the wood ant (*Formica rufa.*) *Br. J. Psychol.*, **56**, 15–31.

270A. VOWLES, D. M. and HARWOOD, D. (1966). The effect of exogenous hormones on aggressive and defensive behaviour in the ring dove (*Streptopelia risoria*). *J. Endocr.*, **36**, 35–51.

271. WALL, W. VON DE (1963). Bewgungsstudien an Anatinen. *J. Orn., Lpz.*, **104**, 1–15.

272. WALSH, E. G. (1964). *Physiology of the Nervous System*, 2nd edn. Longmans, Green, London.

273. WARREN, J. M. (1965). Primate learning in comparative perspective. In *Behavior of Nonhuman Primates*, ed. SCHRIER, A. M., HARLOW, H. F. and STOLLNITZ, F., vol. I, pp. 249–81. Academic Press, New York and London.

273A. WASHBURN, S. L. and DE VORE, I. (1961). Social behavior of baboons and early man. In WASHBURN, S. L. (ed.) *Social Life of Early Man.* pp. 91–105. Aldine Press, Chicago.

274. WATSON, A. J. (1961). The place of reinforcement in the explanation of behaviour. In *Current Problems in Animal Behaviour*, pp. 273–301 (see Ref. 257).

275. WATSON, J. B. (1924). *Behaviorism*. University of Chicago Press, Chicago.

276. WEIDMANN, R. and WEIDMANN, U. (1958). An analysis of the stimulus situation releasing food-begging in the black-headed gull. *Anim. Behav.*, **6**, 114.

277. WEIDMANN, U. (1961). The stimuli eliciting begging in gulls and terns. *Anim. Behav.*, **9**, 115–16.

279. WELLS, M. J. (1958). Factors affecting reactions to *Mysis* by newly hatched *Sepia*. *Behaviour*, **13**, 96–111.

280. WELLS, M. J. (1962). Early learning in *Sepia*. *Symp. zool. Soc. Lond.*, **8**, 149–69.

281. WELLS, M. J. (1962). *Brain and Behaviour in Cephalopods*. Heinemann Educational, London.

281A. WELLS, M. J. (1966). Split brains and octopuses. *Sci. Prog. Oxf.*, **54**, 561–74.

281B. WHITE, S. D., WAYNER, M. J. and COTT, A. (1970). Effects of intensity, water deprivation, prior water ingestion and palatability on drinking evoked by lateral hypothalamic electric stimulation. *Physiol. Behav.*, **5**, 611–19.

281C. WHITTEN, W. K. (1966). Pheromones and mammalian reproduction. *Adv. Reprod. Physiol.*, **1**, 155–77.

281D. WHITTEN, W. K., BRONSON, F. H. and GREENSTEIN, J. A. (1968). Estrus-inducing pheromone of male mice: transport by movement of air. *Science, N.Y.*, **161**, 584–5.

281E. WIENS, J. A. (1966). On group selection and Wynne-Edwards' hypothesis. *Am. Scient.*, **54**, 273–87.

282. WIGGLESWORTH, V. B. (1954). *The Physiology of Insect Metamorphosis.* Cambridge University Press, Cambridge.

283. WILSON, D. M. (1964). The origin of the flight-motor command in grasshoppers. In *Neural Theory and Modeling*, ed. REISS, R. F., pp. 331–45. Stanford University Press, Stanford, California.

283A. WILSON, E. O. (1965). Chemical communication in the social insects. *Science, N.Y.*, **149**, 1064–71.

283B. WILSON, E. O. (1971). *The Insect Societies.* Belknap Press of Harvard University Press, Cambridge, Mass.

283C. WILZ, K. J. (1970). Causal and functional analysis of dorsal pricking and nest activity in the courtship of the three-spined stickleback *Gasterosteus aculeatus*. *Anim. Behav.* **18**, 115–24.

283D. WISE, R. A. (1968). Hypothalamic motivational systems: fixed or plastic neural circuits? *Science, N.Y.*, **162**, 377–79.

283E. WISE, R. A. (1969). Plasticity of hypothalamic motivational systems. *Science, N.Y.*, **165**, 929–30.

284. WOOD-GUSH, D. G. M. (1960). A study of sex drive of two strains of cockerel through three generations. *Anim. Behav.*, **8**, 43–53.

285. WOOD-GUSH, D. G. M. (1971). *The Behaviour of the Domestic Fowl.* Heinemann Educational, London.

286. WOODROW, H. (1942). The problem of general quantitative laws in psychology. *Psychol. Bull.*, **39**, 1–27.

287. WYNNE-EDWARDS, V. C. (1962). *Animal Dispersion in Relation to Social Behaviour.* Oliver and Boyd, Edinburgh.

287A. WYNNE-EDWARDS, V. C. (1964). Group selection and kin selection. *Nature, Lond.*, **201**, 1147.

287B. YOSHIBA, K. (1968). Local and intertroop variability in ecology and social behavior of common Indian langurs. In *Primates: Studies in Adaptation and Variability*, pp. 217–42. (see Ref. 131B).

288. YOUNG, J. Z. (1964). *A Model of the Brain.* Oxford University Press, London.

289. YOUNG, J. Z. (1965). The organization of a memory system. *Proc. R. Soc. B*, **163**, 285–320.

290. YOUNG, W. C., ed. (1961). *Sex and Internal Secretions*, 3rd. edn., 2 vols. Baillière, Tindall and Cox, London.

291. ZANGWILL, O. L. (1961). Lashley's concept of cerebral mass action. In *Current Problems in Animal Behaviour*, pp. 59–86 (see Ref. 257).

292. ZEIGLER, H. P. (1964). Displacement activity and motivational theory. A case study in the history of ethology. *Psychol. Bull.*, **61**, 362–76.

293. ZORNETZER, S. and MCGAUGH, J. L. (1969). Effects of electroconvulsive shock upon inhibitory avoidance: the persistence and stability of amnesia. *Comm. Behav. Biol.*, *A*, **3**, 173–80.

Index

Bold page numbers refer to major entries